AF581764

Atomic Processes in Plasmas and Gases: Symmetries and Beyond

Atomic Processes in Plasmas and Gases: Symmetries and Beyond

Editor

Eugene Oks

MDPI • Basel • Beijing • Wuhan • Barcelona • Belgrade • Manchester • Tokyo • Cluj • Tianjin

Editor
Eugene Oks
Auburn University
USA

Editorial Office
MDPI
St. Alban-Anlage 66
4052 Basel, Switzerland

This is a reprint of articles from the Special Issue published online in the open access journal *Symmetry* (ISSN 2073-8994) (available at: https://www.mdpi.com/journal/symmetry/special_issues/Atomic_Processes_Plasmas_Gases_Symmetries_Beyond).

For citation purposes, cite each article independently as indicated on the article page online and as indicated below:

LastName, A.A.; LastName, B.B.; LastName, C.C. Article Title. *Journal Name* **Year**, *Volume Number*, Page Range.

ISBN 978-3-0365-5443-3 (Hbk)
ISBN 978-3-0365-5444-0 (PDF)

Contents

About the Editor

Eugene Oks

Eugene Oks received his Ph.D. degree from the Moscow Institute of Physics and Technology and, later, the highest degree of Doctor of Sciences from the Institute of General Physics of the Academy of Sciences of the USSR by the decision of the Scientific Council led by the Nobel Prize winner and academician A.M. Prokhorov. According to the Statute of the Doctor of Sciences degree, this highest degree is awarded only to the most outstanding Ph.D. scientists who founded a new research field of a great interest. Oks worked in Moscow (USSR) as the head of a research unit at the Center for Studying Surfaces and Vacuum, then—at the Ruhr University in Bochum (Germany) as an invited professor, and for the last 30 plus years—at the Physics Department of the Auburn University (USA) as a Professor. He conducted research in five areas: atomic and molecular physics, astrophysics, plasma physics, laser physics, and nonlinear dynamics. He founded/co-founded and developed new research fields, such as intra-Stark spectroscopy (a new class of nonlinear optical phenomena in plasmas), masing without inversion (advanced schemes for generating/amplifying coherent microwave radiation), and quantum chaos (nonlinear dynamics in the microscopic world). He also developed a large number of advanced spectroscopic methods for diagnosing various laboratory and astrophysical plasmas—methods that are now used by many experimental groups around the world. He recently revealed that there are two flavors of hydrogen atoms, as proven by the analysis of atomic experiments; there is also a possible astrophysical proof—from observations of the 21 cm radio line from the early Universe. He showed that dark matter, or at least a part of it, can be represented by the second flavor of hydrogen atoms. He published about 550 papers and 11 books, including the books *Plasma Spectroscopy: The Influence of Microwave and Laser Fields*, *Stark Broadening of Hydrogen and Hydrogenlike Spectral Lines in Plasmas: The Physical Insight*, *Breaking Paradigms in Atomic and Molecular Physics*, *Diagnostics of Laboratory and Astrophysical Plasmas Using Spectral Lineshapes of One-, Two-, and Three-Electron Systems*, *Unexpected Similarities of the Universe with Atomic and Molecular Systems: What a Beautiful World*, *Analytical Advances in Quantum and Celestial Mechanics: Separating Rapid and Slow Subsystems*, *Advances in X-ray Spectroscopy of Laser Plasmas*, *Simple Atomic and Molecular Systems: New Results and Applications*, *Advances in the Physics of Rydberg Atoms and Molecules*, and *The Second Flavor of Hydrogen Atoms—the Leading Candidate for Dark Matter: Theoretical Discovery and the Proofs from Experiments and Astrophysical Observations*. He is the Chief Editor of the journal *International Review of Atomic and Molecular Physics* and the Editor-in-Chief of the Physical Sciences section of the journal *Foundations*. He is a member of the Editorial Boards of seven other journals: *Symmetry*, the *American Journal of Astronomy and Astrophysics*, *Dynamics*, *Open Physics*, *Open Journal of Microphysics*, *Physics International*, and *Current Physics*. He is a member of the Reviewers Board of the journal *Atoms*. He is also a member of the International Program Committees of the two series of conferences: Spectral Line Shapes, as well as Zvenigorod Conference on Plasma Physics and Controlled Fusion.

 MDPI

Editorial

Special Issue Editorial "Atomic Processes in Plasmas and Gases: Symmetries and Beyond"

Eugene Oks

Physics Department, Auburn University, 380 Duncan Drive, Auburn, AL 36849, USA; goks@physics.auburn.edu

Atomic processes in plasmas and gases encompass broad areas in theoretical and experimental atomic and molecular physics. One example is atomic processes that are involved in the study of various plasmas over a wide range of electron densities (from 10^{11} cm^{-3} to 10^{23} cm^{-3}) and temperatures (from eV to a few keVs). The topics in this field include (but are not limited to) magnetic fusion plasmas, laser-produced plasmas, relativistic laser–plasma interactions, powerful radiation sources (Z-pinches, plasma focus, XFEL, etc.), low-temperature and industrial plasmas, astrophysical plasmas, as well as plasma spectroscopy for all of the abovementioned applications. Another example is atomic and molecular processes in neutral gases. The topics in this field include (but are not limited to) the molecular spectroscopy of gases, from low-resolution to ultra-high-resolution, from the microwave to the ultraviolet, and from fundamental science to applications such as astronomy and atmospheric science.

Considerations of symmetry often play an important role in theoretical advances, especially in plasma spectroscopy and molecular spectroscopy. For example, the employment of additional conserved quantities, originating from algebraic symmetries of underlying quantum systems, frequently allows important analytical results to be obtained and/or leads to more robust codes.

N.L. Popov and A.V. Vinogradov, in the paper "Space-Time Coupling: Current Concept and Two Examples from Ultrafast Optics Studied Using Exact Solution of EM Equations" [1], discussed the manifestation of space–time coupling (STC) phenomena in the framework of the simplest exact localized solution of Maxwell's equations, exhibiting a "collapsing shell". They considered the excitation of a two-level system located in the center of the collapsing EM (electromagnetic) pulse. This study showed that as it propagates, a unipolar pulse can turn into a bipolar one, and in the case of measuring the excitation efficiency, we can judge which of these two pulses we are dealing with. The obtained results have no limitation on the number of cycles in a pulse. The work confirmed the productivity of using exact solutions of EM wave equations for describing the phenomena associated with STC effects.

M. Goto and N. Ramaiya, in the paper "Polarization of Lyman-α Line Due to the Anisotropy of Electron Collisions in a Plasma" [2], developed an atomic model for the calculation of the polarization state of the Lyman-α line in plasma caused by anisotropic electron collision excitations. The calculation results gave the polarization degree of several percent under typical conditions in the edge region of a magnetically confined fusion plasma. They also found that the relaxation of polarization due to collisional averaging among the magnetic sublevels was effective in the electron density region considered. Their analysis of the experimental data measured in the Large Helical Device yielded $T\perp/T\|$ = 7.6 at the expected Lyman-α emission location outside the confined region. The result was derived with the absolute polarization degree of 0.033, and $T\perp$ = 32 eV and ne = 9.6 × 1018 m^{-3}, measured using the Thomson scattering diagnostic system.

P.A. Sdvizhenskii et al., in the paper "Data for Beryllium–Hydrogen Charge Exchange in One and Two Centres Models, Relevant for Tokamak Plasmas" [3], presented the analysis of data on the cross section and kinetic rate of charge exchange (CX) between the bare beryllium nucleus, the ion Be(+4), and the neutral hydrogen atom. These data are of

Citation: Oks, E. Special Issue Editorial "Atomic Processes in Plasmas and Gases: Symmetries and Beyond". *Symmetry* **2022**, *14*, 1497. https://doi.org/10.3390/sym14081497

Received: 24 June 2022
Accepted: 18 July 2022
Published: 22 July 2022

great interest for visible-range high-resolution spectroscopy in the ITER tokamak because beryllium is intended as the material for the first wall in the main chamber. Data in the range of a few eV/amu to ~100 eV/amu (amu stands for the atomic mass unit) needed for simulations of level populations for principal and orbital quantum numbers in the emitting beryllium ions Be(+3) can be obtained with the help of two-dimensional kinetic codes. The lack of literature data, especially for data resolved in orbital quantum numbers, has prompted us to make numerical calculations with the ARSENY code. The authors presented the comparison of the results obtained for the one-center Coulomb problem using an analytic approach with the two-center problem using numerical simulations.

C.G. Parigger et al., in the paper "Hypersonic Imaging and Emission Spectroscopy of Hydrogen and Cyanide Following Laser-Induced Optical Breakdown" [4], communicated the connection of measured shadowgraphs from optically induced air breakdown with emission spectroscopy in selected gas mixtures. Laser-induced optical breakdown was generated using 850 and 170 mJ and 6 ns pulses at a wavelength of 1064 nm, the shadowgraphs were recorded using time-delayed 5 ns pulses at a wavelength of 532 nm and a digital camera, and emission spectra were recorded for typically a dozen discrete time-delays from optical breakdown by employing an intensified charge-coupled device. The symmetry of the breakdown event could be viewed as close-to spherical symmetry for time-delays of several 100 ns. The analysis of the air breakdown and selected gas breakdown events permitted the use of Abel inversion for the inference of the expanding species distribution. Overall, the recorded air breakdown shadowgraphs were indicative of laser–plasma expansion in selected gas mixtures, and optical spectroscopy delivered analytical insight into plasma expansion phenomena.

V.A. Astapenko and E.V. Sakhno, in the paper "Chirped Laser Pulse Effect on a Quantum Linear Oscillator" [5], presented a theoretical study of the excitation of a charged quantum linear oscillator via a chirped laser pulse by using the probability of the process throughout the pulse action. They focused on the case of the excitation of the oscillator from the ground state without relaxation. Calculations were made for an arbitrary value of the electric field strength by utilizing the exact expression for the excitation probability. The dependence of the excitation probability on the pulse parameters was analyzed both numerically and by using analytical formulas.

E. Oks, in the paper "Oscillatory-Precessional Motion of a Rydberg Electron Around a Polar Molecule" [6], provided a detailed classical description of the oscillatory–precessional motion of an electron in the field of an electric dipole. Specifically, he demonstrated that in the general case of the oscillatory–precessional motion of an electron (with the oscillations being in the meridional direction (θ-direction) and the precession being along parallels of latitude (φ-direction)), both the θ-oscillations and the φ-precessions can actually occur on the same time scale—contrary to the statement from the work by another author. He obtained the dependence of φ on θ, the time evolution of the dynamical variable θ, the period T_θ of the θ-oscillations, and the change in the angular variable φ during one half-period of the θ-motion—all in the forms of one-fold integrals in the general case—and illustrated it pictorially. The author also produced the corresponding explicit analytical expressions for relatively small values of the projection p_φ of the angular momentum on the axis of the electric dipole. He also derived a general condition for this conditionally periodic motion to become periodic (the trajectory of the electron would become a closed curve) and then provide examples of the values of p_φ for this to happen. In addition, for the particular case of $p_\varphi = 0$, he produced an explicit analytical result for the dependence of the time t on θ. For the opposite particular case, where p_φ is equal to its maximum possible value (consistent with the bound motion), he derived an explicit analytical result for the period of the revolution of the electron along the parallel of latitude.

E. Oks, in the paper "Application of the Generalized Hamiltonian Dynamics to Spherical Harmonic Oscillators" [7], extended the applications of the Dirac's Generalized Hamiltonian Dynamics (GHD) to a charged Spherical Harmonic Oscillator (SHO). Dirac's Generalized Hamiltonian Dynamics (GHD) is a purely classical formalism for systems having constraints: it incorporates the constraints into the Hamiltonian. Dirac designed the GHD specifically for applications to quantum field theory. In one of Oks' previous papers (coauthored with T. Uzer) [8], he redesigned Dirac's GHD for its applications to atomic and molecular physics by choosing integrals of the motion as the constraints. In that paper, after a general description of the formalism, they considered hydrogenic atoms as an example. They showed that this formalism leads to the existence of classical non-radiating (stationary) states and that there is an infinite number of such states—just as in the corresponding quantum solution. In the present paper, while extending the applications of the GHD to the SHO, Oks demonstrated that, by using the higher-than-geometrical symmetry (i.e., the algebraic symmetry) of the SHO and the corresponding additional conserved quantities, it is possible to obtain the classical non-radiating (stationary) states of the SHO and that, generally speaking, there is an infinite number of such states of the SHO. Both the existence of the classical stationary states of the SHO and the infinite number of such states are consistent with the corresponding quantum results. He obtained these new results from first principles. Physically, the existence of the classical stationary states is the manifestation of a non-Einsteinian time dilation. Time dilates more and more as the energy of the system becomes closer and closer to the energy of the classical non-radiating state. He emphasized that the SHO and hydrogenic atoms are not the only microscopic systems that can be successfully treated by the GHD. All classical systems of N degrees of freedom have the algebraic symmetries O_{N+1} and SU_N, and this does not depend on the functional form of the Hamiltonian. In particular, all classical spherically symmetric potentials have algebraic symmetries, namely O_4 and SU_3; they possess an additional vector integral of the motion, while the quantal counterpart-operator does not exist. This offers possibilities that are absent in quantum mechanics.

Conflicts of Interest: The author declares no conflict of interest.

References

1. Popov, N.L.; Vinogradov, A.V. Space-Time Coupling: Current Concept and Two Examples from Ultrafast Optics Studied Using Exact Solution of EM Equations. *Symmetry* **2021**, *13*, 529. [CrossRef]
2. Goto, M.; Ramaiya, N. Polarization of Lyman-α Line Due to the Anisotropy of Electron Collisions in a Plasma. *Symmetry* **2021**, *13*, 297. [CrossRef]
3. Sdvizhenskii, P.A.; Tolstikhina, I.Y.; Lisitsa, V.S.; Kukushkin, A.B.; Tugarinov, S.N. Data for Beryllium–Hydrogen Charge Exchange in One and Two Centres Models, Relevant for Tokamak Plasmas. *Symmetry* **2021**, *13*, 16. [CrossRef]
4. Parigger, C.G.; Helstern, C.M.; Gautam, G. Hypersonic Imaging and Emission Spectroscopy of Hydrogen and Cyanide Following Laser-Induced Optical Breakdown. *Symmetry* **2021**, *12*, 2116. [CrossRef]
5. Astapenko, V.A.; Sakhno, E.V. Chirped Laser Pulse Effect on a Quantum Linear Oscillator. *Symmetry* **2021**, *12*, 1293. [CrossRef]
6. Oks, E. Oscillatory-Precessional Motion of a Rydberg Electron Around a Polar Molecule. *Symmetry* **2021**, *12*, 1275. [CrossRef]
7. Oks, E. Application of the Generalized Hamiltonian Dynamics to Spherical Harmonic Oscillators. *Symmetry* **2021**, *12*, 1130. [CrossRef]
8. Oks, E.; Uzer, T. Application of Dirac's Generalized Hamiltonian Dynamics to Atomic and Molecular Systems. *J. Phys. B At. Mol. Opt. Phys.* **2002**, *35*, 165–173. [CrossRef]

Article

Space-Time Coupling: Current Concept and Two Examples from Ultrafast Optics Studied Using Exact Solution of EM Equations

Nikolay L. Popov * and Alexander V. Vinogradov

P.N. Lebedev Physical Institute, Leninsky Prospekt 53, 119991 Moscow, Russia; vinograd@sci.lebedev.ru
* Correspondence: popovnl@sci.lebedev.ru

Abstract: Current approach to space-time coupling (STC) phenomena is given together with a complementary version of the STC concept that emphasizes the finiteness of the energy of the considered pulses. Manifestations of STC are discussed in the framework of the simplest exact localized solution of Maxwell's equations, exhibiting a "collapsing shell". It falls onto the center, continuously deforming, and then, having reached maximum compression, expands back without losing energy. Analytical solutions describing this process enable to fully characterize the field in space-time. It allowed to express energy density in the center of collapse in the terms of total pulse energy, frequency and spectral width in the far zone. The change of the pulse shape while travelling from one point to another is important for coherent control of quantum systems. We considered the excitation of a two-level system located in the center of the collapsing EM (electromagnetic) pulse. The result is again expressed through the parameters of the incident pulse. This study showed that as it propagates, a unipolar pulse can turn into a bipolar one, and in the case of measuring the excitation efficiency, we can judge which of these two pulses we are dealing with. The obtained results have no limitation on the number of cycles in a pulse. Our work confirms the productivity of using exact solutions of EM wave equations for describing the phenomena associated with STC effects. This is facilitated by rapid progress in the search for new types of such solutions.

Keywords: space-time couplings; spatiotemporal; ultrafast optics; unipolar pulses; few cycle pulses

Citation: Popov, N.L.; Vinogradov, A.V. Space-Time Coupling: Current Concept and Two Examples from Ultrafast Optics Studied Using Exact Solution of EM Equations. *Symmetry* **2021**, *13*, 529. https://doi.org/10.3390/sym13040529

Academic Editor: Eugene Oks

Received: 1 March 2021
Accepted: 22 March 2021
Published: 24 March 2021

1. Introduction

Emergence and development of ultrashort laser pulses [1,2] and ultrafast optics technology [3] stimulated the interest of researchers to laser pulses with the duration equal to few, one and even less periods of electromagnetic field [4–7]. Production, characterization and manipulation with these pulses are required by many applications in various branches of physics, chemistry, biology and medicine [8,9]. At the very beginning of the era of ultra-fast optics, it turned out that temporal pulse shaping changes its spatial spectrum (see for example [10] and references herein). In other words, it is not possible to control a beam in space and time independently. This is a manifestation of a linear optical effect, which is present not only in a dispersive medium, but also in empty space, and is called space-time coupling (STC).

STC is a fundamental property of real coherent EM beams. However, for pulses containing many periods of the field, STC is not important and is difficult to observe, whereas it is very important for few cycle pulses. Due to the wide range of research on ultrafast optics numerous papers, books and tutorials are devoted to STC. They often employ quite different and conflicting approaches to explanation and simulation of STC phenomenon.

This paper discusses two issues in which STC plays an important role. The first is what is the maximum energy density achieved in the center of a collapsing EM beam? The second: What is the efficiency of energy transfer of such a beam to a two state quantum mechanical system placed in the center? As was already mentioned above, STC is a very general property and therefore we use exact solutions of EM wave equations as a natural and reliable basis for STC theory and simulation. The results are valid for pulses of any duration from quasi-monochromatic to sub-period.

2. Materials and Methods

2.1. What Does the Concept of Spatio-Temporal Couplings Mean in the Theory of EM Waves?

Google Scholar indexes about twenty thousand papers on spatio-temporal couplings in EM pulses.

However, there is no generally accepted definition of the term (Wikipedia does not contain the article on this subject) and various authors explain it in different ways, for example: (a) in the language of formulas, it is said that the electric field of a beam cannot be represented by the product of functions of spatial and temporal coordinates [11] and (b) a beam whose temporal or spectral properties depend on space, or vice versa, is said to exhibit spatio-temporal couplings [12].

Our STC concept is closer to (b) and can be briefly formulated as follows: Any real pulse, that is, having a finite energy and obeying the equations of electromagnetic waves, has a temporal form that depends on coordinates, and its spatial forms are constantly changing. Such coupled variability is as fundamental property of EM pulses as the conservation laws of classical invariants: Energy, momentum, angular momentum, spin and Zeldovich invariants [13–17].

The most obvious consequence of STC is transformation of the temporal shape during travelling of a few cycle or subcycle beam. This is important in such problems as achievement of extreme laser field intensities [18] and manipulation of quantum systems with laser radiation [19,20]. It is easy to be convinced that plane waves and Gaussian beams cannot be used to simulate STC in these (and other) problems, as their energy is infinite. Fourier optics and superpositions of plane waves or Gaussian beams can be used. This certainly complicates the modeling [10].

At the same time, in the last three decades, the study of analytical solutions of Maxwell's equations describing localized electromagnetic waves began, also largely due to the needs of ultrafast optics. They can be found in [4,13,18,21–25].

As is known [26,27], any EM wave can be constructed from linear transformations of solutions of the scalar wave equation:

$$\left(\Delta - \frac{1}{c^2}\frac{\partial^2}{\partial t^2}\right)u(\boldsymbol{r},t) = 0. \tag{1}$$

However, formulas for exact solutions of Equation (1), as well as for Maxwell's equations, are often rather complicated. Their study can be an independent task.

Nevertheless, the scalar wave equation allows us to explain the STC principle. In the case of spherical symmetry, Equation (1) takes the form:

$$\frac{1}{r^2}\frac{\partial}{\partial r}r^2\frac{\partial u}{\partial r} - \frac{1}{c^2}\frac{\partial^2 u}{\partial t^2} = 0. \tag{2}$$

Its solutions are traveling spherical waves [28]:

$$\frac{f(ct+r)}{r} \text{ and } \frac{f(ct-r)}{r}, \tag{3}$$

where $f(x)$ is an arbitrary, fast-decreasing function. Their linear combination

$$u(r,t) = \frac{f(ct+r) - f(ct-r)}{r}, \; r \geq 0, \; -\infty < t < \infty \tag{4}$$

has no singularity at $r = 0$ [29,30] and can be used to construct the exact solution of Maxwell's equations [24,31].

As is seen the scalar wave (4) describes a collapsing spherical shell, which is falling onto the center $r = 0$ and then expanding back. This is a direct demonstration of the STC principal. Indeed, firstly, the variables r and t cannot be separated in two factors, with the product equal to $u(r,t)$. Then, at each point r, while t changes from $-\infty$ to $+\infty$, the wave $u(r,t)$ is continuously transformed from an incoming to an outcoming (reflected from the center) wave. As a result, the temporal shape strongly depends on the radius r, giving at the center ($r=0$):

$$u(r=0,t) = 2f\prime(ct). \tag{5}$$

For example, if $f(x) = exp\left\{-\frac{x^2}{a^2}\right\}$, the pulse shape $u(r,t)$ is a sequence of two bunches, incident (positive) and reflected (negative), separated at $r > a$ by an interval

$$T \approx \frac{2r}{c}\left(1 + 2exp\left\{-4\left(\frac{r}{a}\right)^2\right\}\right). \tag{6}$$

If the radius is reduced, the shape of both bunches will change, and at the center $r=0$ the pulse (4) is transformed into (5). Then the interval is reduced to $T = \sqrt{2}\frac{a}{c}$. The above scenario is illustrated in Figure 1 and is generally preserved for a vector field, but the calculations become more cumbersome. In the next Section, following [24,31], we present the results for the field of the simplest electromagnetic pulse corresponding to (4).

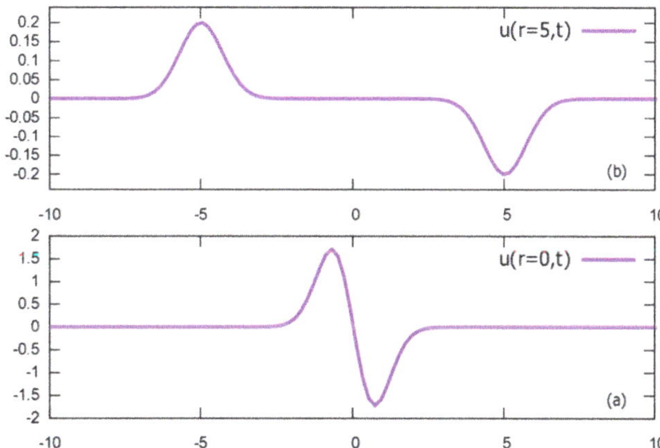

Figure 1. (**a**) is the function $u(r=0,t)$ and (**b**) is the function $u(r-5,t)$, if $f(x) = exp(\ x^2)$, $a=1$. Horizontally, the value of ct, vertically—u in arbitrary units.

2.2. Exact Solutions of Free Space Electromagnetic Wave Equations with Finite Total Energy

Electromagnetic wave constructed from arbitrary solution $u(\mathbf{r},t)$ of scalar wave Equation (1) has the following field structure (elementary derivation of (7) is given in [24,31] if we make the replacement $\mathbf{E} \to \mathbf{H}$, $\mathbf{H} \to -\mathbf{E}$, corresponding to duality transformation):

$$\begin{aligned} \mathbf{E}(\mathbf{r},t) &= -\mathbf{l}\Delta u(\mathbf{r},t) + (\mathbf{l}\nabla)\nabla u(\mathbf{r},t), \\ \mathbf{H}(\mathbf{r},t) &= -\frac{1}{c}\frac{\partial}{\partial t}\mathbf{l} \times \nabla u(\mathbf{r},t), \end{aligned} \tag{7}$$

where $\mathbf{l}$ is an arbitrary axial unit vector. If $u(\mathbf{r},t) = u(r,t)$ is spherically symmetric and has the form (4), expressions (7) demonstrate a collapsing vector electromagnetic pulse, which is very similar to a collapsing scalar shell, considered in the previous section. They are greatly simplified in near and far zones. In the far zone, when $ct \to -\infty$ and $r = x - ct \to \infty$ (x is of the order of the pulse length) the pulse has the form of incoming wave falling onto the center:

$$\mathbf{E} \approx \frac{\mathbf{n}(\mathbf{l}\mathbf{n}) - \mathbf{l}}{r} g(ct+r), \tag{8}$$

$$\mathbf{H} \approx \frac{\mathbf{n} \times \mathbf{l}}{r} g(ct+r), \tag{9}$$

where

$$\boldsymbol{n} = \frac{\boldsymbol{r}}{r}, \ g(x) = f''(x). \tag{10}$$

In the center of the collapse zone, at $r = 0$ the electric and magnetic wavefields are deduced from (7) and give:

$$\boldsymbol{E}(t) = -\frac{4}{3}\boldsymbol{l}g'(ct), \tag{11}$$

$$\boldsymbol{H} = 0. \tag{12}$$

Comparison of the fields in the near (11), (12) and far (8), (9) zones convincingly demonstrates STC. First, in the center, in contrast to (8), (9), only the electric field is nonzero and, secondly, its time dependence does not coincide with the shape of the incident beam and is transformed into its time derivative.

Knowing the fields, one can use the general formulas [27] to determine the electromagnetic energy density $\epsilon(\boldsymbol{r}, t)$ and then the total pulse energy $\boldsymbol{\mathcal{E}}$ [22,24,31]:

$$\boldsymbol{\mathcal{E}} = \int \epsilon(\boldsymbol{r}, t)\, d\boldsymbol{r} = \frac{2}{3}\int_{-\infty}^{\infty} g^2(s)ds, \tag{13}$$

which, naturally, does not depend on time. The convergence of integral (13) is related to the assumption of rapid decay of the function f (x) (see (3) and (4)), which determines $g(x)$ (see (8)–(10)).

Explicit expressions for the field of a light wave (7)–(12) allow comparing their characteristics in different regions of space. Namely, to predict the results of measuring the spectrum and shape of pulses at different positions of the detector. To do this, it is necessary to agree on the exact definitions of the compared quantities in the frequency and time domains.

Electric and magnetic fields in classical physics are real quantities. Therefore, the functions g, f and all others proportional to them are real. In this case, physical intuition suggests the following natural definitions of pulse duration, frequency and line width [32]:

$$\text{pulse duration–}\sigma_t = \sqrt{\langle t^2 \rangle - \langle t^2 \rangle}, \tag{14}$$

$$\text{where } \langle t \rangle = \frac{\int_{-\infty}^{+\infty} t g^2(t)dt}{\int_{-\infty}^{+\infty} g^2(t)dt}, \langle t^2 \rangle = \frac{\int_{-\infty}^{+\infty} t^2 g^2(t)dt}{\int_{-\infty}^{+\infty} g^2(t)dt}; \tag{15}$$

$$\text{average frequency–}\omega_0 = \langle \omega \rangle = \frac{\int_0^{+\infty} \omega |G(\omega)|^2 d\omega}{\int_0^{+\infty} |G(\omega)|^2 d\omega}, \tag{16}$$

$$\text{where } G(\omega) = \int_{-\infty}^{+\infty} g(t)\exp(-i\omega t)dt; \tag{17}$$

$$\text{line width–}\sigma_\omega = \sqrt{\langle \omega^2 \rangle - \omega_0^2}, \tag{18}$$

$$\text{where } \langle \omega^2 \rangle = \frac{\int_0^{+\infty} \omega^2 |G(\omega)|^2 d\omega}{\int_0^{+\infty} |G(\omega)|^2 d\omega}. \tag{19}$$

Note that (14)–(19) differ from the usual definitions of the moments of the complex Fourier transform used in quantum mechanics, and leading to the Heisenberg uncertainty relation. The difference is that 0 appears instead of $-\infty$ as the lower frequency integration limit. As noted above, this is due to the reality of fields in classical physics. To establish the uncertainty relation, it was necessary for the first time in physics to turn to the properties of complex numbers (this circumstance was noted by D. Gabor [33]). In more detail than in this article, it is discussed in [31,32].

Thus, the field of a collapsing pulse (7), which is an exact solution of Maxwell's equations, contains an arbitrary function $g(x)$, through which the total pulse energy and characteristics of its spectrum are expressed. This, along with accurate temporal and spectral characterization (14)–(19), ensures consistent accounting of STC effects.

The next two sections use a Sinusoidal–Gaussian (the word sinusoidal will be omitted in the rest of this article for brevity) function as the incident wave:

$$g(x) = Ce^{-\frac{x^2}{a^2}} sinqx. \tag{20}$$

As will be seen further coefficients C, a and q can be expressed in terms of pulse total energy and spectrum. This makes it possible to consider in a unified manner quasi-monochromatic and ultrashort pulses, including one- and sub-period.

2.3. *Maximum Energy Density of a Collapsing EM Beam*

Exact analytical pulse-like solutions of Maxwell's equations are helpful for practical estimations and benchmarking the results of computer simulation as well as approximate models of EM beams widely used in many branches of physics, chemistry, biology and other fields. If we are talking about a few cycles and, often, femtosecond pulses, then STC can play a significant role. Evidently exact solutions of EM equations provide a rigorous approach to their accounting. This is demanded by various problems of ultrafast optics including formation of superstrong laser fields, for studying quantum electrodynamic effects, the use of femtosecond pulses in nonlinear optics, focusing, manipulation and control of laser radiation in the presence of spatiotemporal couplings, etc. (see [8,12,34] and references therein).

The above applies to the collapsing EM beam described by (7) and (4), especially since the latter contains an arbitrary function $f(x)$. It allows to estimate maximum energy density achievable with given pulse energy $\boldsymbol{\mathcal{E}}$. The energy density at the center can be easily obtained using (11), (12):

$$\epsilon(\boldsymbol{r},t)\Big|_{r=0} = \frac{1}{8\pi}\Big[\boldsymbol{E}^2(\boldsymbol{r},t) + \boldsymbol{H}^2(\boldsymbol{r},t)\Big]\Big|_{r=0} = \frac{2}{9\pi}\left[g'(ct)\right]^2. \tag{21}$$

Then dividing (21) by total pulse energy (13) yields:

$$\frac{\epsilon(\boldsymbol{r},t)|_{r=0}}{\mathcal{E}} = \frac{[g'(x)]^2}{3\pi\int_{-\infty}^{+\infty} g^2(x)dx}, \quad x = ct. \tag{22}$$

Formula (22) as well as (21) and (13) is valid for arbitrary incident pulse shape function $g(x)$. For a Gaussian pulse shape (20) many expressions are simplified. Total energy is found by direct integration in (13):

$$\mathcal{E} = aC^2\frac{\sqrt{2\pi}}{6}\left(1 - e^{-\frac{a^2q^2}{2}}\right). \tag{23}$$

Analysis of the time dependence for $g(x)$ from (20) using Formula (21) shows that the maximum flux density at the center of the collapse is:

$$\epsilon_m = Max[\epsilon\,(\boldsymbol{r},t)|_{r=0}] = \left(\frac{2}{\pi}\right)^{\frac{3}{2}}\mathcal{E}\frac{q^2}{3a}\left(1 - e^{-\frac{a^2q^2}{2}}\right)^{-1}. \tag{24}$$

So, the energy density ϵ_m produced in the center of a collapsing EM pulse is proportional to the total pulse energy $\mathcal{E}$ and depends on temporal shape of the incident pulse through parameters $\frac{1}{q}$ and a, having the dimension of length. Formula (24) allows us to talk about laser beam engineering in the far zone to achieve the maximum field in the center [22].

For a quasi-monochromatic beam (with many periods $N = aq$ of the light field), the Gaussian parameters a and q are evidently reduced to the central beam frequency and linewidth:

$$\omega_0 = qc; \ \sigma_\omega = \frac{c}{a}, \quad N \gg 1. \tag{25}$$

Then expression (24) takes on a simple and understandable form [31]:

$$\frac{\epsilon_m}{\mathcal{E}} = \frac{8}{3}\left(\frac{\sqrt{2\pi}}{\lambda_0}\right)^3 \frac{1}{N}; \text{or } \frac{\epsilon_m}{\mathcal{E}} = \frac{8}{3}\left(\frac{\sqrt{2\pi}}{\lambda_0}\right)^3 \frac{\sigma_\omega}{\omega_0}, \ \frac{\omega_0}{\sigma_\omega} = N \gg 1, \tag{26}$$

where $\lambda_0 = \frac{2\pi c}{\omega_0}$ is the central wavelength. Formula (26) claims that in ultrashort pulses, with a decrease in the number of periods, an increasing fraction of the energy is concentrated in the volume $\sim (\lambda_0)^3$. For few periods and sub-period pulses (small N), the situation must be analyzed on the basis of consistent application of definitions (14)–(19). This approach is implemented in the works [24,31].

For arbitrary number of time periods N, the maximum available flux density ϵ_m as well as formulas for pulse shape parameters in time and frequency domains are given in the Appendix A.

2.4. Collapsing Shell: "Conventional or Strange Wave"?

In this Section, we will consider another property of the collapsing shell, which, as it turns out, goes beyond waves in free space. E.G. Bessonov in his works [35,36] published 40 and 30 years ago showed that the electric field vector of the far zone radiation produced by a charge moving in a finite space region satisfies condition (we use the letter $\boldsymbol{S}$ instead of his $\boldsymbol{I}$ for consistency with the works of subsequent authors):

$$\boldsymbol{S}(\boldsymbol{r}) = \int_{-\infty}^{\infty} \boldsymbol{E}(\boldsymbol{r}, t)dt = 0. \tag{27}$$

Based on (27), he concluded that a bounded charge system cannot be a source of unipolar (single sign) waves and suggested using the value $\boldsymbol{S}(\boldsymbol{r})$ for classification of electromagnetic waves and description their properties. The waves obeying (27) he called "conventional" and the waves with $\boldsymbol{S}(\boldsymbol{r}) \neq \boldsymbol{0}$ (including unipolar waves) were called strange waves. Then in the same papers he considered several elementary processes to generate strange waves. This problem, especially in relation to unipolar pulses, is of fundamental and applied interest. The works [35,36] received a noticeable response in accelerator and microwave communities. Since then, several theoretical and experimental papers on e-beam and other sources of strange and unipolar waves have been published [37–45]. Interest in the topic increased sharply in the late 90s. Generation, application and study of unipolar pulses has become extremely relevant with the advent of the era of few cycle laser fields. The main findings of [35] were again analyzed and confirmed [46,47]. The parameter S is now also used in a broader sense than the criteria given by Bessonov's Formula (27). For characterization of strange waves which are not unipolar the authors of [48] introduced the degree of unipolarity:

$$\xi(\boldsymbol{r}) = \frac{|\boldsymbol{S}(\boldsymbol{r})|}{\int_{-\infty}^{+\infty} |\boldsymbol{E}(\boldsymbol{r})| dt} = \frac{\left|\int_{-\infty}^{+\infty} \boldsymbol{E}(\boldsymbol{r}) dt\right|}{\int_{-\infty}^{+\infty} |\boldsymbol{E}(\boldsymbol{r})| dt}. \tag{28}$$

Let us determine the place of the collapsing electromagnetic pulse in the above classification. Before answering the question posed in the title of the section, we note that the above-mentioned Bessonov's works considered the radiation of a moving charge—Lienard–Wiechert potentials—the radiation field of a charge or a system of charges.

Nevertheless, it turns out that the relation $\boldsymbol{S}(\boldsymbol{r}) = 0$ is also fulfilled for an EM pulse collapsing in a vacuum. This is easy to verify by integrating $\boldsymbol{E}(\boldsymbol{r}, t)$ in (7) over time. It leads, in accordance with (4), to the subtraction of two identical integrals. Therefore, the collapsing beam for which $u(\boldsymbol{r},t)$ is a spherically symmetric function (4) also satisfies (27). So, in the above classification, this is a conventional wave.

To conclude this Section, there are two more remarks concerning Formula (27). First, it is also valid for the magnetic field $\boldsymbol{H}(\boldsymbol{r},t)$ of the collapsing spherical shell. Indeed, since $\boldsymbol{H}(\boldsymbol{r},t)$ in (7) is a time derivative, it is easy to see that the integral in (27) for $\boldsymbol{H}(\boldsymbol{r},t)$, given by (7), vanishes for an arbitrary solution $u(\boldsymbol{r},t)$ of the scalar wave Equation (1).

Secondly, according to Whittaker's theorem "Only two solutions of the scalar wave equation are needed to represent an arbitrary electromagnetic field in empty space" [26,27]. The fields $\boldsymbol{E}(\boldsymbol{r},t)$ and $\boldsymbol{H}(\boldsymbol{r},t)$ are obtained as a result of the action of linear differential operators of the first order on these solutions. Hence it follows that if the solutions of the scalar wave equation that appear under the conditions of Whittaker's theorem, or are used to construct the exact solution of Maxwell's equations, are spherically symmetric, then the corresponding fields $\boldsymbol{E}(\boldsymbol{r},t)$ and $\boldsymbol{H}(\boldsymbol{r},t)$ satisfy condition (27). Finally, A.B. Plachenov showed that the condition (27) is satisfied by the fields of an arbitrary electromagnetic pulse if they decrease sufficiently rapidly in space and time [49]. A detailed consideration of this issue is beyond the scope of our work.

2.5. *Excitation of a Two-Level Atom Placed at the Center of a Collapsing Beam*

Returning to the principles of STC, discussed in Section 2.1, we can say that the manifestation of STC is a continuous change in the shape and spectrum of a propagating EM pulse. A consistent description of this effect is given by exact solutions of Maxwell's equations corresponding to finite total pulse energy. The key words here are "finite total energy". It is in this case that it is possible to unambiguously relate the characteristics of the EM pulse in different regions of space. For example, Formulas (24) and (26) relate the maximum energy density at the center of the collapse with the temporal shape of the pulse and its spectrum in the far zone.

As another instructive example, consider the efficiency of excitation of an atom by an EM finite-energy pulse. This formulation of the question is encountered in the problems of manipulating atoms with laser radiation that arise in chemistry, quantum optics, physics of trapped atoms and ions, trace elements and other fields [19]. The subtlety lies in the fact that the shape and spectrum of an incident pulse with a finite total energy in the far zone differs from that arriving at the point where the atom is located. This manifestation of the STC effect is again elegantly accounted for with the asymptotic expressions (8), (9) and (11), (12) of exact solution (7), (4) to Maxwell's equations.

Let a two-level atom located at the center of the collapsing beam at $r = 0$ be described by the probability amplitudes in the ground $a_1(t)$ and excited $a_2(t)$ states, so that the wave function of the atom has the form:

$$\boldsymbol{\Psi}(t) = e^{-\frac{i}{\hbar}E_1 t}a_1(t)\boldsymbol{\psi}_1 + e^{-\frac{i}{\hbar}E_2 t}a_2(t)\boldsymbol{\psi}_2, \tag{29}$$

where E_1, $\boldsymbol{\psi}_1$ and E_2, $\boldsymbol{\psi}_2$ are the energies and wave functions of the system in the ground and excited states, respectively. Then the Schrödinger equation for the wave function (29) is reduced to a system of ordinary differential equations [19,50]:

$$\begin{cases} i\hbar\dot{a}_1 = V(t)e^{-i\omega t}a_2 \\ i\hbar\dot{a}_2 = V^*(t)e^{i\omega t}a_1 \end{cases}, \quad \hbar\omega = E_2 - E_1, \tag{30}$$

where $V(t)$ is the off-diagonal matrix element of the perturbation associated with the field of the incident electromagnetic pulse, which we take in the form:

$$V(t) = -d * E(t), \tag{31}$$

where d is the dipole moment of the atom, and $E(t)$ is the field of the EM pulse at the point where the atom is.

Since this article deals with pulses with a finite total energy, it is clear, that

$$V(t) \to 0, \text{ for } t \to \pm\infty. \tag{32}$$

Assuming that at $t \to -\infty$, the atom is in the ground state, let us consider the probability of its excitation by an EM pulse at $t \to +\infty$. To do this, obviously, it is necessary to solve the system (30) with the initial conditions:

$$\begin{cases} a_1(-\infty) = 1 \\ a_2(-\infty) = 0 \end{cases} \tag{33}$$

and calculate $a_2(+\infty)$. Excitation efficiency η is the ratio of the energy acquired by the atom

$$\mathcal{E}_a = \hbar\omega|a_2(+\infty)|^2, \tag{34}$$

to the total energy of the incident pulse $\mathcal{E}$:

$$\eta(g) = \frac{\mathcal{E}_a}{\mathcal{E}} = \frac{\hbar\omega|a_2(+\infty)|^2}{\mathcal{E}}. \tag{35}$$

The value of $\eta(g)$ naturally depends on the shape of the incident pulse $g(x)$, since both the numerator and denominator in (35), in accordance with (30), (31), (11) and (13), are defined in terms of $g\ (x)$.

Thus, to find the excitation efficiency $\eta(g)$, it is necessary to solve the system of equations (30) with the initial conditions (33). The explicit analytical solution of (30) is known only for several specific functions $V(t)$ and is described by rather cumbersome expressions [19]. Therefore, in practice, one should focus on the numerical solution of problem (30) and (33). However, in the case of a weak field, $E(t)$ the perturbation theory is valid:

$$a_2(t) \approx -\frac{i}{\hbar}\int_{-\infty}^{t} V(t\prime)e^{i\omega t\prime}dt\prime \tag{36}$$

and

$$\hbar^2|a_2(+\infty)|^2 \approx \left|\int_{-\infty}^{\infty} V(t)e^{i\omega t}dt\right|^2 = d^2\left|\int_{-\infty}^{\infty} E(t)e^{i\omega t}dt\right|^2. \tag{37}$$

Hence it follows that in the case of a small incident pulse energy of a collapsing pulse the excitation efficiency of an atom located in the center takes the form:

$$\eta(g) = \frac{8}{3}\frac{d^2\omega^3}{\hbar c^2}\frac{\left|\int_{-\infty}^{\infty} g(t)e^{i\omega t}dt\right|^2}{\int_{-\infty}^{\infty} g^2(s)ds},\ s = ct. \tag{38}$$

Formula (38) is obtained by substituting (11) into (31) and then into (36) and (35). It is seen that the value $\eta(g)$ does not depend on the total pulse energy. It is it that can be considered a small parameter. Indeed, in accordance with (23) and (20), the total pulse energy determines the scale of the magnitude of the electromagnetic field, which is related to the perturbation theory used in deriving (37).

As in Section 2.3 (see (26)) for a Gaussian pulse (20), the excitation efficiency can be expressed in terms of the spectral parameters ω_0 иσ_ω of the incident pulse.

Accurate accounting of STC reveals some interesting effects for coherent pulse incident onto a quantum system. Firstly, the incident unipolar (or according to [5] half period) pulse:

$$g(x) = Ce^{-\frac{x^2}{a^2}},\ x = ct, \tag{39}$$

after propagation to the center, as follows from Section 3 (see (11)), is transformed into one period [5] or, in terms of Section 2.2, Formula (27), conventional pulse:

$$E(r=0,t) \sim g'(ct) = C_1 x e^{-\frac{x^2}{a^2}},\ x = ct. \tag{40}$$

Second, the presence of a quantum system at the focal point can in principle be used to unambiguously establish whether a pulse is one-period ($S = 0$ as in (27)) or half-period ($S \neq 0$, i.e., unipolar). To make sure of this, consider the transition amplitude (36) for small values of the transition frequency ω:

$$a_2(+\infty,\omega) \approx -\frac{i}{\hbar}\int_{-\infty}^{\infty} V(t)e^{i\omega t}dt = -\frac{i}{\hbar}\int_{-\infty}^{\infty} V(t)dt + \frac{\omega}{\hbar}\int_{-\infty}^{\infty} tV(t)dt + \cdots, \tag{41}$$

where $V(t)$ is a perturbation proportional to the field $E(t)$ acting on an atom. Accordingly, for the transition probability we obtain:

$$w_{1\to 2}(\omega) = |a_2(+\infty,\omega)|^2 = \frac{d^2S^2}{\hbar^2} + \frac{d^2\omega^2}{\hbar^2}\left|\int_{-\infty}^{\infty} tE(t)dt\right|^2 + \cdots,\ S = \int_{-\infty}^{\infty} E(t)dt. \tag{42}$$

Thus, as seen from (42), the dependence of the transition probability on the resonance defect ω for unipolar ($S \neq 0$) and conventional ($S = 0$) pulses is fundamentally different. This difference makes it possible to judge the structure of ultrashort laser pulses. For experimental verification and usage, the quantum state engineering of trapped atomic particles developed in recent decades [51–53] can be proposed.

3. Results

(a) The concept of space-time couplings of electromagnetic pulses is complemented by the important requirement of finiteness of total pulse energy.
(b) The field of a collapsing electromagnetic beam is found in space and time basing on the exact solution of Maxwell's equations in terms of the total energy, the spectrum and number of cycles in the incident pulse.
(c) The excitation efficiency of a two-level quantum system placed in the center of a collapsing beam is found with a full account for space-time couplings.
(d) The analysis showed that electromagnetic field distributions originated by solutions of scalar wave equation cannot be single sign (unipolar).
(e) The method to experimentally distinguish between conventional and unipolar pulses is suggested.

4. Conclusions

STCs are usually associated with electromagnetic fields, which are described by functions with nonseparated spatial (x, y, z) and temporal (t) coordinates. In other words, fields that cannot be represented as the product of coordinate and temporal factors. At the same time, the opinion is often met that for practical purposes this does not matter. The latter is confirmed by wide application of plane waves, Gaussian beams, as well as fields in the form of products of the spatial and temporal (low- or sub-period) parts in various problems of laser and atomic physics, optics, including imaging and ultrafast optics. However, these approximations are not directly suitable for STC modeling. It is necessary to use functions with nonseparated spatial and temporal coordinates. Hope for the possibility of efficient and rigorous accounting of STC effects is given by exact analytical solutions of free space Maxwell's equations. Their search and study are intensively developed after the works of R. W. Ziolkowski [54,55]. In this case, an important condition is the finiteness of the total pulse energy. A review and recent references on this subject are in [21,56,57].

In addition to the above, we consider STC as a definite property of any EM pulse possessing finite energy. This property is as fundamental, as the conservation laws of classical invariants: Total energy, momentum, angular momentum, spin and Zeldovich invariants [13–17]. Here it is appropriate to mention, following Bessonov, the formula of "conventional waves" (27), which says: Each of the projections of the electromagnetic field of a finite energy pulse at any point in space is a sign-variable function of time, the integral of which is zero.

The evident manifestation of STC is a continuous change in the shape and spectrum of a propagating EM pulse. This means that the space field distribution changes in time and vice versa: The observed pulse shape and spectrum change from point to point.

In this regard, if we consider the EM impulse as a material object, it is interesting to point out the discussion that has been going on for many years [58,59] around the painting "Rain, Steam and Speed—The Great Western Railway" (1844) by the brilliant English artist J.W.M. Turner, see Figure 2. It depicts [60] a steam locomotive—the fastest vehicle at the time—on the newly opened railway. "The feeling of speed is conveyed by the darker color of the locomotive in relation to the surrounding space, in which no object has a clear outline. Turner was almost not interested in the forms of the miracle of technology—the locomotive with its now seemingly old-fashioned tall pipe, he wanted to convey the movement ... Lindsay [58], subtly noticed that the rapid movement of the locomotive is conveyed by the fact that it is made darker and clearer than anything else. *The difference between them expresses the sequence of movement in time*" [59]. Now we can assume, that depicting the most powerful and perfect technical creation in the picture, Turner, perhaps, expressed his presentiment of the physical picture of the structure of matter opening to humanity, part of which is a propagating electromagnetic pulse of finite energy, obeying the then still unknown Maxwell's equations.

Figure 2. "Rain, Steam and Speed—The Great Western Railway" (1844), J.W.M. Turner.

Our concept of STC naturally leads us in Sections 2.3 and 2.5 to quantitative approach to analysis of STC effects based on exact solutions to Maxwell's equations and accurate definitions of spectral parameters of real signals. This allows a unified description of quasi-monochromatic, few period and sub-period pulses.

In Section 2.3 and Appendix A, this approach is used for a detailed quantitative analysis of the structure of a collapsing electromagnetic shell. At the moment of collapse, in the center it takes the form of a ball, which, flying apart, again turns into a shell. Rigorous consideration of STC effects allows expressing the field at the center, including the maximum EM energy density, in terms of the frequency and linewidth of the incident radiation. In this way, it is possible to select the shape of the laser pulse in order to achieve the desired behavior of the electromagnetic field in the center.

In Section 2.5, taking STC effects into account, the efficiency of energy transfer of a collapsing electromagnetic pulse to a two-level atom located in the center is determined. In this case, the differences in the shape and spectrum of the pulse in the far and near zones can be especially important. For example, an incident half-period pulse transforms into a one-period pulse at the center, which significantly changes the excitation efficiency for a small energy level difference. This result opens the possibility of practically distinguishing between conventional and unipolar (strange) electromagnetic pulses.

Thus, there is reason to expect that the use of exact solutions of Maxwell's equations can become an efficient, rigorous approach to the study of subtle and complex phenomena from STC to laser action on atoms and polyatomic objects. The further development of this direction will be facilitated by the search for more sophisticated and realistic exact solutions that correspond to the finite total pulse energy and thereby bring us closer to experiment.

The authors are indebted to N.V. Dyachkov, A.A. Gorbatsevich, V.I. Man'ko, A.B. Plachenov, I.E. Protsenko, I.V. Smetanin for valuable discussions and E.A. Rakhmanov, who took part in writing the Appendix A.

Author Contributions: Both authors contributed equally. All authors have read and agreed to the published version of the manuscript.

Funding: This research received no external funding.

Institutional Review Board Statement: Not applicable.

Informed Consent Statement: Not applicable.

Data Availability Statement: Not applicable.

Conflicts of Interest: The authors declare no conflict of interest.

Appendix A. Spectral Analysis of a Gaussian Pulse (20) Based on a "Real Signal" Definitions (14)–(19)

D. Gabor [33] and then I. Kay and R.F. Silverman [32] pointed out that the theory of the Fourier transform of complex functions and the associated Heisenberg uncertainty relation cannot be applied to the spectral analysis of classical signals that are real functions. This circumstance is critically important for ultrashort laser pulses, when the spectrum width and average frequency are comparable in magnitude. At the same time, for narrow-band signals, the effect does not manifest itself, and simpler formulas for the complex Fourier transform can be used as a rigorous approximation.

For reference, we give general formulas for pulse shape parameters in frequency (ω_0, σ_ω) and time ($\langle t\rangle$, σ_t) domains calculated according to (14)–(19) for Gaussian function (20):

$$\omega_0 = \frac{c}{a}\frac{N}{1-e^{-\frac{N^2}{2}}} erf\left(\frac{N}{\sqrt{2}}\right), \tag{A1}$$

$$\langle\omega^2\rangle = \frac{c^2}{a^2}\frac{1-e^{\frac{N^2}{2}}(1+N^2)}{1-e^{\frac{N^2}{2}}}, \tag{A2}$$

$$\sigma_\omega = \frac{c}{a}\sqrt{\frac{e^{\frac{N^2}{2}}(1+N^2)-1}{e^{\frac{N^2}{2}}-1} - \frac{N^2}{\left(1-e^{-\frac{N^2}{2}}\right)^2}\mathrm{erf}^2\left(\frac{N}{\sqrt{2}}\right)}, \tag{A3}$$

$$\langle t\rangle = 0, \tag{A4}$$

$$\langle t^2\rangle = \frac{a^2}{4c^2}\frac{N^2+e^{\frac{N^2}{2}}-1}{e^{\frac{N^2}{2}}-1}, \tag{A5}$$

$$\sigma_t = \frac{a}{2c}\sqrt{\frac{N^2+e^{\frac{N^2}{2}}-1}{e^{\frac{N^2}{2}}-1}}. \tag{A6}$$

Formulas (A1)–(A6) are valid for arbitrary number N of electromagnetic field periods.

For quasi-monochromatic pulses when $N = aq \gg 1$ Formulas (A1)–(A6) easily yield (25):

$$\omega_0 = qc = \frac{cN}{a};\ \sigma_\omega = \frac{c}{a};\ \langle t^2 \rangle = \frac{a^2}{4c^2};\ \sigma_t = \frac{a}{2c}, \tag{A7}$$

whereas for sub-period pulses when $N = aq \ll 1$ one obtains:

$$\omega_0 = \frac{c}{a}\frac{2\sqrt{2}}{\sqrt{\pi}};\ \sigma_\omega = \frac{c}{a}\sqrt{3 - \frac{8}{\pi}};\ \langle t^2 \rangle = \frac{3a^2}{4c^2};\ \sigma_t = \frac{a}{c}\frac{\sqrt{3}}{2} \tag{A8}$$

in agreement with [31].

Now we can analyze the results of Section 2.3 in terms of physically significant and measurable quantities: Center frequency ω_0 and line width σ_ω (or, when convenient, N).

For example, the ratio of maximum flux density to the total pulse energy for an arbitrary number of field periods obviously follows from (23), (24) and (A1):

$$\frac{\epsilon_m}{\mathcal{E}} = \frac{q^3}{3N}\frac{1}{1 - e^{-\frac{N^2}{2}}}\left(\frac{2}{\pi}\right)^{\frac{3}{2}} = \frac{(8\pi)^{\frac{3}{2}}}{3N\lambda_0^3}\frac{\left(1 - e^{-\frac{1}{2}N^2}\right)^2}{\left[erf\left(\frac{N}{\sqrt{2}}\right)\right]^3}, \tag{A9}$$

where $\lambda_0 = \frac{2\pi c}{\omega_0}$. For large number of periods $N = aq \gg 1$ this yields (26).

References

1. Brabec, T.; Krausz, F. Intense few-cycle laser fields: Frontiers of nonlinear optics. *Rev. Mod. Phys.* **2000**, *72*, 545. [CrossRef]
2. Diels, J.; Rudolph, W. *Ultrashort Laser Pulse Phenomena*, 2nd ed.; Academic Press: Cambridge, MA, USA, 2006.
3. Weiner, A.M. Ultrafast optical pulse shaping: A tutorial review. *Opt. Commun.* **2011**, *284*, 3669–3692. [CrossRef]
4. Feng, S.; Winful, H.G.; Hellwarth, R.W. Spatiotemporal evolution of focused single-cycle electromagnetic pulses. *Phys. Rev. E* **1999**, *59*, 4630. [CrossRef]
5. Keldysh, L.V. Multiphoton ionization by a very short pulse. *Phys. Usp.* **2017**, *60*, 1187–1193. [CrossRef]
6. Park, S.B.; Kim, K.; Cho, W.; Hwang, S.I.; Ivanov, I.; Nam, C.H.; Kim, K.T. Direct sampling of a light wave in air. *Optica* **2018**, *5*, 402–408. [CrossRef]
7. Hwang, S.I.; Park, S.B.; Mu, J.; Cho, W.; Nam, C.H.; Kim, K.T. Generation of a single-cycle pulse using a two-stage compressor and its temporal characterization using a tunnelling ionization method. *Sci. Rep.* **2019**, *9*, 1613. [CrossRef] [PubMed]
8. Nolte, S.; Schrempel, F.; Dausinger, F. *Ultrashort Pusle Laser Technology*; Springer International Publishing: New York, NY, USA, 2016.
9. Sun, H.; Fritz, A.; Dröge, G.; Neuhann, T.; Bille, J.F. Femtosecond-Laser-Assisted Cataract Surgery (FLACS). In *High Resolution Imaging in Microscopy and Ophthalmology*; Bille, J.F., Ed.; Springer: Cham, Switzerland, 2019; pp. 301–317.
10. Frei, F.; Galler, A.; Feurer, T. Space-time coupling in femtosecond pulse shaping and its effects on coherent control. *Chem. Phys.* **2009**, *130*, 034302. [CrossRef]
11. Jolly, S.W.; Gobert, O.; Quéré, F. Spatio-temporal characterization of ultrashort laser beams: A tutorial. *Optica* **2017**, *4*, 1298–1304. [CrossRef]
12. Sainte-Marie, A.; Gobert, O.; Quéré, F. Controlling the velocity of ultrashort light pulses in vacuum through spatio-temporal couplings. *Optica* **2017**, *4*, 1298–1304. [CrossRef]
13. Lekner, J. *Theory of Electromagnetic Pulses*; Morgan & Claypool Publishers: San Rafael, CA, USA, 2018.
14. Zeldovich, Y.B. Number of quanta as an invariant of the classical electromagnetic field. *Sov. Phys. Dokl.* **1966**, *10*, 771.
15. Bialynicki-Birula, I. Photon Wave Number. In *Progress in Optics*; Wolf, E., Ed.; Elsevier Science: New York, NY, USA, 1996; Volume 36, pp. 245–294.
16. Feshchenko, R.M.; Vinogradov, A.V. On the number of photons in a classical electromagnetic field. *J. Exp. Theor. Phys.* **2018**, *127*, 274–278. [CrossRef]
17. Feshchenko, R.M.; Vinogradov, A.V. On the number and spin of photons in classical electromagnetic fields. *Phys. Scr.* **2019**, *94*, 065501. [CrossRef]
18. Fedotov, A.M.; Korolev, K.Y.; Legkov, M.V. Exact analytical expression for the electromagnetic field in a focused laser beam or pulse. *Proc. SPIE* **2007**, *6726*, 672613.
19. Shore, B.W. *Manipulating of Quantum Structures with Laser Pulses*; Cambridge University Press: Cambridge, UK, 2011.
20. Vitanov, N.V.; Shore, B.W. Designer evolution of quantum systems by inverse engineering. *J. Phys. B At. Mol. Opt.* **2016**, *48*, 174008. [CrossRef]
21. Kiselev, A.P. Localized Light Waves: Paraxial and Exact Solutions of the Wave Equation (a Review). *Opt. Spectrosc.* **2007**, *102*, 603–622. [CrossRef]

22. Gonoskov, I.; Aiello, A.; Heugel, S.; Leuchs, G. Dipole pulse theory: Maximizing the field amplitude from 4π focused laser pulses. *Phys. Rew. A* **2012**, *86*, 053836. [CrossRef]
23. Hernández-Figueroa, H.E.; Zamboni-Rached, M.; Recami, E. *Non-Diffracting Waves*; John Wiley & Sons: New York, NY, USA, 2013.
24. Artyukov, I.A.; Dyachkov, N.V.; Feshchenko, R.M.; Vinogradov, A.V. Energy density and spectrum of single-cycle and sub-cycle electromagnetic pulses. *Quantum Electron.* **2020**, *50*, 187–194. [CrossRef]
25. So, I.A.; Plachenov, A.B.; Kiselev, A.P. Unidirectional Single-Cycle and Sub-Cycle Pulses. *Opt. Spectrosc.* **2020**, *128*, 2005–2006. [CrossRef]
26. Bateman, H. *The Mathematical Analysis of Electrical and Optical Wave-motion on the Basis of Maxwell's Equations*; Cambridge University Press: Cambridge, UK, 1915.
27. Zangwill, A. *Modern Electrodynamics*; Cambridge University Press: Cambridge, UK, 2012.
28. Feynman, R.P.; Leighton, R.B.; Sands, M.L. Solutions of Maxwell's Equations in Free Space. In *The Feynman Lectures on Physics*; Addison-Wesley Pub. Co.: Boston, MA, USA, 1964; Volume 2, Chapter 20; p. 4.
29. Tikhonov, A.N.; Samarskii, A.A. *Equations of Mathematical Physics*; Dover Publications: New York, NY, USA, 1963.
30. Landau, L.D.; Lifshitz, E.M. Spherical Waves. In *Fluid Mechanics*, 2nd ed.; Pergamon Press: Oxford, UK, 1987; p. 70.
31. Artyukov, I.A.; Dyachkov, N.V.; Feshchenko, R.M.; Vinogradov, A.V. Collapsing EM wave—A simple model for nonparaxial, quasimonochromatic, single and half-cycle beams. *Phys. Scr.* **2020**, *95*, 064006. [CrossRef]
32. Kay, I.; Silverman, R.A. On the Uncertainty Relation for Real Signals. *Inf. Control* **1957**, *1*, 64–75. [CrossRef]
33. Gabor, D. Theory of communication. Part 1: The analysis of information. *J. Inst. Elec. Eng.* **1946**, *93*, 429–441. [CrossRef]
34. Froula, D.H.; Palastro, J.P.; Turnbull, D.; Davies, A.; Nguyen, L.; Howard, A.; Ramsey, D.; Franke, P.; Bahk, S.-W.; Begishev, I.A.; et al. Flying focus: Spatial and temporal control of intensity for laser-based applications. *Phys. Plasmas* **2019**, *26*, 032109. [CrossRef]
35. Bessonov, E.G. On a class of electromagnetic waves. *Zh. Eksp. Teor. Fiz.* **1981**, *80*, 852–858.
36. Bessonov, E.G. Conventionally strange electromagnetic waves. *Nucl. Instr. Meth. A* **1991**, *308*, 135–139. [CrossRef]
37. Shibata, Y.; Bessonov, E.G. Long Wavelength Broadband Sources of Coherent Radiation. *arXiv* **1997**, arXiv:physics/9708023.
38. Bratman, V.L.; Jaroszynski, D.A.; Samsonov, S.V.; Savilov, A.V. Generation of ultra-short quasi-unipolar electromagnetic pulses from quasi-planar electron bunches. *Nucl. Instr. Meth. A* **2001**, *475*, 436–440. [CrossRef]
39. Alexeev, V.I.; Bessonov, E.G. Experiments on the generation of long wavelength edge radiation along directions nearly coincident with the axis of a straight section of the "Pakhra" synchrotron. *NIM* **2001**, *173*, 54–60. [CrossRef]
40. Schwarz, M.; Basler, P.; Borstel, M.V.; Müller, A.S. Analytic calculation of the electric field of a coherent THz pulse. *Phys. Rev. Spec. Top. Accel. Beams* **2014**, *17*, 050701. [CrossRef]
41. Balal, N.; Bratman, V.L.; Savilov, A.V. Peculiarities of the coherent spontaneous synchrotron radiation of dense electron bunches. *Phys. Plasmas* **2014**, *21*, 023103. [CrossRef]
42. Freund, F.T.; Herau, J.A.; Centa, V.A.; Scoville, J. Mechanism of unipolar electromagnetic pulses emitted from the hypocenters of impending earthquakes. *Eur. Phys. J. Spec. Top.* **2021**, *230*, 47–65. [CrossRef]
43. Fedorov, V.M.; Ostashev, V.E.; Tarakanov, V.P.; Ul'yanov, A.V. High power radiators of ultra-short electromagnetic quasi-unipolar pulses. *J. Phys. Conf. Ser.* **2017**, *830*, 012020. [CrossRef]
44. Naumenko, G.; Shevelev, M. First indication of the coherent unipolar diffraction radiation generated by relativistic electrons. *JINST* **2018**, *13*, C05001. [CrossRef]
45. Naumenko, G., Shevelev, M., Popov, K.E. Unipolar Cherenkov and Diffraction Radiation of Relativistic Electrons. *Phys. Part Nucl. Lett.* **2020**, *17*, 834–839. [CrossRef]
46. Kim, K.J.; McDonald, K.T.; Stupakov, G.V.; Zolotorev, M.S. A bounded source cannot emit a unipolar electromagnetic wave. *arXiv* **2000**, arXiv:physics/0003064.
47. Kim, K.J.; McDonald, K.T.; Stupakov, G.V.; Zolotorev, M.S. Comment on "Coherent Acceleration by Subcycle Laser Pulses". *Phys. Rev. Lett.* **2000**, *84*, 3210. [CrossRef] [PubMed]
48. Arkhipov, R.M.; Pakhomov, A.V.; Arkhipov, M.V.; Babushkin, I.; Tolmachev, Y.A.; Rosanov, N.N. Generation of unipolar pulses in nonlinear media. *JETP Lett.* **2017**, *105*, 408–418. [CrossRef]
49. Plachenov, A.B. Paraxial beams and related solutions of the Helmholtz equation. In Proceedings of the International Conference DAYS ON DIFFRACTION, St. Petersburg, Russia, 25–29 May 2020.
50. Landau, L.D.; Lifshitz, E.M. *Quantum Mechanics*, 3rd ed.; Pergamon Press: Oxford, UK, 1958.
51. Jun, Y.; Kimble, H.J.; Katori, H. Quantum State Engineering and Precision Metrology Using State-Insensitive Light Traps. *Science* **2008**, *320*, 1734–1738.
52. Cho, D.; Hong, S.; Lee, M.; Kim, T. A review of silicon microfabricated ion traps for quantum information processing. *Micro Nano Syst. Lett.* **2015**, *3*, 2. [CrossRef]
53. Zhang, Q.; Wang, Y.; Zhu, C.; Wang, Y.; Zhang, X.; Gao, K.; Zhang, W. Precision measurements with cold atoms and trapped ions. *Chin. Phys. B* **2020**, *29*, 093203. [CrossRef]
54. Ziolkowski, R.W. Exact solutions of the wave equation with complex source locations. *J. Math. Phys.* **1985**, *26*, 861. [CrossRef]
55. Ziolkowski, R.W. Localized Waves: Historical and Personal Perspectives. In *Non-Diffracting Waves*; Hernández-Figueroa, H.E., Recami, E., Eds.; John Wiley & Sons: New York, NY, USA, 2013; Chapter 2, see [23].
56. So, I.A.; Plachenov, A.B.; Kiselev, A.P. Simple unidirectional finite-energy pulses. *Phys. Rev. A* **2020**, *102*, 063529. [CrossRef]

57. Zdagkas, A.; Papasimakis, N.; Savinov, V. Space-time nonseparable pulses: Constructing isodiffracting donut pulses from plane waves and single-cycle pulses. *Phys. Rev. A* **2020**, *102*, 063512. [CrossRef]
58. Lindsay, J. *J.M.W. Turner. His Life and Work;* Cary, Adams & Mackay: London, UK, 1966.
59. Nekrasova, E.A. Turner 1775–1851. Moscow, Russia. 1976. Available online: https://www.amazon.com/Nekrasova-Terner-1775-1851-Nekrasov-Turner/dp/B0718YYFZY (accessed on 22 March 2021).
60. The National Gallery. Available online: https://www.nationalgallery.org.uk/paintings/joseph-mallord-william-turner-rain-steam-and-speed-the-great-western-railway (accessed on 22 March 2021).

 *symmetry* 

Article

Polarization of Lyman-α Line Due to the Anisotropy of Electron Collisions in a Plasma

Motoshi Goto [1,2,*] and Nilam Ramaiya [3]

1 National Institute for Fusion Science, Toki 509-5292, Japan
2 Department of Fusion Science, The Graduate University for Advanced Studies, SOKENDAI, Toki 509-5292, Japan
3 Institute for Plasma Research, Gandhinagar 382428, India; nilam@ipr.res.in
* Correspondence: goto.motoshi@nifs.ac.jp

Abstract: We have developed an atomic model for calculating the polarization state of the Lyman-α line in plasma caused by anisotropic electron collision excitations. The model assumes a nonequilibrium state of the electron temperature between the directions parallel ($T_{\parallel}$) and perpendicular ($T_{\perp}$) to the magnetic field. A simplified assumption on the formation of an excited state population in the model is justified by detailed analysis of population flows regarding the upper state of the Lyman-α transition with the help of collisional-radiative model calculations. Calculation results give the polarization degree of several percent under typical conditions in the edge region of a magnetically confined fusion plasma. It is also found that the relaxation of polarization due to collisional averaging among the magnetic sublevels is effective in the electron density region considered. An analysis of the experimental data measured in the Large Helical Device gives $T_{\perp}/T_{\parallel} = 7.6$ at the expected Lyman-α emission location outside the confined region. The result is derived with the absolute polarization degree of 0.033, and $T_{\perp} = 32\,\text{eV}$ and $n_e = 9.6 \times 10^{18}\,\text{m}^{-3}$ measured by the Thomson scattering diagnostic system.

Keywords: plasma spectroscopy; polarization; Lyman-alpha; nuclear fusion

Citation: Goto, M.; Ramaiya, N. Polarization of Lyman-α Line Due to the Anisotropy of Electron Collisions in a Plasma. *Symmetry* **2021**, *13*, 297. https://doi.org/10.3390/sym13020297

Academic Editor: Eugene Oks
Received: 22 January 2021
Accepted: 5 February 2021
Published: 9 February 2021

1. Introduction

In a magnetically confined fusion plasma, the velocity distribution function (VDF) of electrons and ions is thought to be more or less anisotropic. For example, energetic ions are unidirectionally introduced by the neutral beam, and the cyclotron motions of ions and electrons are selectively accelerated by the electron cyclotron resonance heating (ECRH) and the ion cyclotron resonance heating (ICRH), respectively. Because of the unavoidable magnetic field ripple, confinement characteristics are different between the particles having a large pitch angle and a small pitch angle with respect to the magnetic field. The former and the latter are called trapped particles and passing particles, respectively, and are sometimes treated separately when the particle transport is considered.

An example of the research relating to this topic is the influence of plasma pressure anisotropy on the MHD (Magnetohydrodynamic) equilibria, which has been intensively investigated [1]. The ITER experiment would also be influenced by the problem of anisotropy [2]. In the Large Helical Device (LHD), the so-called density clamping observed with strong ECRH is thought to be attributed to the difference in the confinement characteristics of the trapped and passing electrons [3]. The anisotropy also plays a role in the plasma edge region. The radial electric field formation in the edge stochastic region is thought to be related to the anisotropy in the electron VDF (EVDF) [4].

Although the anisotropy is regarded as an important subject for characterizing the plasma confinement as seen above, no reliable measurement method for the anisotropic VDF of electrons and ions has been established to date. Under such circumstances, polarization spectroscopy has been proposed as a technique to address the problem of anisotropy in the EVDF [5]. This novel diagnostic method consists of two issues, i.e., the measurement of polarization in line emission from the plasma and the construction of an atomic model for analyzing the observation data. A critical problem in the measurement is a difficulty in detecting the polarization of line emission with accuracy, the degree of which is estimated in the order of one percent or smaller. Although some measurements have been attempted, no reliable results have been obtained [6,7].

Recently, the CLASP (Chromospheric Lyman-Alpha SpectroPolarimetry) project led by the NAOJ (National Astronomical Observatory of Japan) has successfully measured the polarization state of the hydrogen Lyman-α line in the solar atmosphere [8]. Although the polarization formation mechanism in the solar atmosphere is photoexcitation by the anisotropic radiation field, which is different from the mechanism in the fusion plasma, the observation technique itself can be transferred to the measurement for a fusion plasma. Actually, the same technique as CLASP has been attempted in LHD, and the polarization of the Lyman-α line has been successfully detected [9].

As for the atomic model, a sophisticated framework has been developed by Fujimoto [5], and some actual applications have been made [6,10]. We here report details of the implementation of Fujimoto's framework for the Lyman-α line with goal of utilizing it for the analysis of the polarization data taken in the LHD experiment. Furthermore, an analysis has been made of the LHD experimental data and derivation of an anisotropic EVDF is attempted.

2. Theoretical Model for the Line Emission Polarization

2.1. Polarization Formation of the Lyman-α Line

The polarization of an emission line originates in a population imbalance between the magnetic sublevels in the upper state of the transition. The Lyman-α line consists of two fine structure lines, i.e., $1\,^2S_{1/2} - 2\,^2P_{1/2}$ and $1\,^2S_{1/2} - 2\,^2P_{3/2}$. Figure 1 shows all transitions between magnetic sublevels composing the Lyman-α line. The $\Delta m_J = 0$ and $\Delta m_J = \pm 1$ transitions emit light linearly polarized in the quantization axis direction (π light) and circularly polarized on the plane perpendicular to the quantization axis (σ light), respectively, where m_J is the magnetic quantum number. The numbers next to the lines indicating the transitions are the relative values of the Einstein A coefficients. It is confirmed that if all the magnetic sublevels have the same population, the intensities of π, σ^+, and σ^- lights are identical, i.e., there is no polarization, and otherwise the line is polarized.

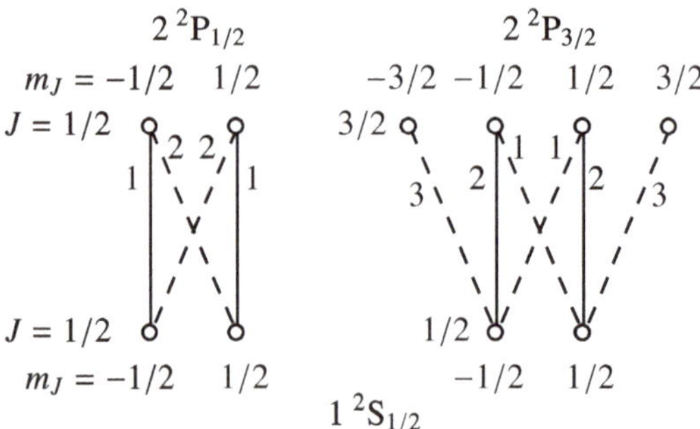

Figure 1. Line components included in the Lyman-α line. The solid and dashed lines represent the π- and σ-light, respectively. The numbers next to the lines indicate relative values of the Einstein A coefficient.

We assume axisymmetry with respect to the quantization axis which is taken in the magnetic field direction later. In this case, the population distribution has a "mirror symmetry", i.e., the populations of m_J and $-m_J$ sublevels are identical in each of the $2\,^2P_{1/2}$ and $2\,^2P_{3/2}$ states. Because of this restriction, the line corresponding to the $1\,^2S_{1/2} - 2\,^2P_{1/2}$

transition is never polarized because π, σ^+, and σ^- light intensities are always identical. As for the $1\,^2\mathrm{S}_{1/2}-2\,^2\mathrm{P}_{3/2}$ transition, populations of the $m_J = |1/2|$ and $m_J = |3/2|$ substates in the $2\,^2\mathrm{P}_{3/2}$ state can be different, which could give rise to the line polarization because only the $m_J = |1/2|$ substates are responsible for the π light. We first focus on deriving the polarization state of the $1\,^2\mathrm{S}_{1/2}-2\,^2\mathrm{P}_{3/2}$ line and then incorporate the influence of the unpolarized $1\,^2\mathrm{S}_{1/2}-2\,^2\mathrm{P}_{1/2}$ line into the result as explained below to enable a direct comparison of the model results with the observation results.

When the measurement is made from the direction perpendicular to the quantization axis with a linear polarizer, the polarization degree P of an emission line is generally defined as

$$P = \frac{I_\pi - I_\sigma}{I_\pi + I_\sigma}, \tag{1}$$

where I_π and I_σ represent the line intensities observed when the polarizer is directed in the direction parallel and perpendicular to the quantization axis, respectively. Because the Lyman-α line includes the two fine structure components, P can be explicitly written as

$$P = \frac{(I_\pi(3/2) + I_\pi(1/2)) - (I_\sigma(3/2) + I_\sigma(1/2))}{(I_\pi(3/2) + I_\pi(1/2)) + (I_\sigma(3/2) + I_\sigma(1/2))}, \tag{2}$$

where $I_\pi(1/2)$ and $I_\sigma(1/2)$ are the intensities of the π- and σ-components of the $1\,^2\mathrm{S}_{1/2}-2\,^2\mathrm{P}_{1/2}$ line, respectively, and $I_\pi(3/2)$ and $I_\sigma(3/2)$ are the same but of the $1\,^2\mathrm{S}_{1/2}-2\,^2\mathrm{P}_{3/2}$ line, respectively. Because the $1\,^2\mathrm{S}_{1/2}-2\,^2\mathrm{P}_{1/2}$ line is unpolarized under the present condition, the relation

$$I_\pi(1/2) - I_\sigma(1/2) = 0 \tag{3}$$

should always hold. On the other hand, we assume that the population ratio of the $2\,^2\mathrm{P}_{1/2}$ state to the $2\,^2\mathrm{P}_{3/2}$ state follows the ratio of their statistical weights, i.e., the former is half the latter. In that case, the same is true for the line intensities of the $1\,^2\mathrm{S}_{1/2}-2\,^2\mathrm{P}_{1/2}$ and $1\,^2\mathrm{S}_{1/2}-2\,^2\mathrm{P}_{3/2}$ transitions, i.e.,

$$I_\pi(1/2) + I_\sigma(1/2) = \frac{1}{2}[I_\pi(3/2) + I_\sigma(3/2)]. \tag{4}$$

By using the relations of Equations (3) and (4), we can rewrite Equation (2) as

$$\begin{aligned} P &= \frac{(I_\pi(3/2) - I_\sigma(3/2)) + (I_\pi(1/2) - I_\sigma(1/2))}{(I_\pi(3/2) + I_\sigma(3/2)) + (I_\pi(1/2) + I_\sigma(1/2))} \\ &= \frac{I_\pi(3/2) - I_\sigma(3/2)}{\frac{3}{2}[I_\pi(3/2) + I_\sigma(3/2)]} \\ &= \frac{2}{3}P(3/2), \end{aligned} \tag{5}$$

where $P(3/2)$ is the polarization degree of the $1\,^2\mathrm{S}_{1/2}-2\,^2\mathrm{P}_{3/2}$ line.

2.2. Polarization Due to Anisotropic Electron Collisions

The condition of an excited state which has a population imbalance among the magnetic sublevels is well represented by the density matrix [11]. Under an axisymmetric system, the spherical coordinate representation of the density matrix ρ of the state $2\,^2\mathrm{P}_{3/2}$ can be expanded [5] as

$$\rho(p) = \rho_0^0(p)T_0^{(0)}(p) + \rho_0^2(p)T_0^{(2)}(p), \tag{6}$$

where p stands for the state $2\,^2\mathrm{P}_{3/2}$ and $T_q^{(k)}(p)$ is the so-called irreducible tensor operator. The coefficients $\rho_0^0(p)$ and $\rho_0^2(p)$ respectively correspond to the population and the alignment, the latter of which expresses the inhomogeneity over the magnetic sublevels in the state p. We hereafter use $a(p)$ instead of $\rho_0^2(p)$ for simplicity. The conventional population is given as $n(p) = \sqrt{2J+1}\rho_0^0(p)$, where J is the total angular momentum quantum number of the state p.

The population imbalance among the magnetic sublevels could be created by anisotropic electron collisions. We consider a simple atomic model for a quantitative calculation of $n(p)$ and $a(p)$ where the population inflow and outflow concerning the state p are balanced by the electron impact excitation of the ground state atoms and the spontaneous radiative decay. Such a condition can be expressed [5] as

$$C^{0,0}(1,p)n_{\rm e}n(1) = \sum_s A(p,s)n(p), \tag{7}$$

where $C^{0,0}(1,p)$ is the rate coefficient of the electron impact excitation from the ground state denoted as "1" to the state p, $A(p,s)$ is the Einstein A coefficient of the transition from p to a state s placed energetically lower than p, and $n_{\rm e}$ is the electron density. The validity of this model under the conditions assumed here will be examined later.

The equilibrium condition of $a(p)$ is similarly expressed [5] as

$$C^{0,2}(1,p)n_{\rm e}n(1) = \left[\sum_s A(p,s) + C^{2,2}(p,p)n_{\rm e}\right]a(p), \tag{8}$$

where $C^{0,2}(1,p)$ is the alignment creation rate coefficient accompanying the excitation from the ground state to the state p and $C^{2,2}(p,p)$ is the alignment destruction rate coefficient in the state p. The population $n(p)$ and the alignment $a(p)$ are then expressed as

$$n(p) = \frac{C^{0,0}(1,p)n_{\rm e}}{\sum_s A(p,s)}n(1), \tag{9}$$

$$a(p) = \frac{C^{0,2}(1,p)n_{\rm e}}{\sum_s A(p,s) + C^{2,2}(p,p)n_{\rm e}}n(1). \tag{10}$$

As discussed in Section 2.1, the measurement gives the polarization degree P. We here introduce another quantity, "longitudinal alignment", $A_{\rm L}$, which is defined slightly differently from P [5] as

$$\begin{aligned} A_{\rm L} &= \frac{I_\pi - I_\sigma}{I_\pi + 2I_\sigma} \\ &= \frac{2P}{3-P}. \end{aligned} \tag{11}$$

The longitudinal alignment for the transition from the state p to s, i.e., $A_{\rm L}(p,s)$, is directly related to $a(p)/n(p)$, the normalized alignment, [5] as

$$A_{\rm L}(p,s) = (-1)^{J_p+J_s}\sqrt{\frac{3}{2}}(2J_p+1)\left\{\begin{array}{ccc} J_p & J_p & 2 \\ 1 & 1 & J_s \end{array}\right\}\frac{a(p)}{n(p)}, \tag{12}$$

where $\{\cdots\}$ is the 6-j symbol, and J_p and J_s are the total angular momentum quantum numbers of the states p and s, respectively.

Calculations of the coefficients $C^{0,0}(1,p)$ and $C^{0,2}(1,p)$ can be carried out under a certain EVDF. We assume that the EVDF is axisymmetric with respect to the quantization axis and is expressed by two temperatures, $T_\parallel$ and $T_\perp$, which are in the directions parallel and perpendicular, respectively, to the quantization axis (or the magnetic field). Such EVDFs are explicitly given [5] as

$$f(v,\theta) = \left(\frac{m}{2\pi k}\right)^{3/2}\left(\frac{1}{T_\perp^2 T_\parallel}\right)^{1/2}\exp\left[-\frac{mv^2}{2k}\left(\frac{\sin^2\theta}{T_\perp} + \frac{\cos^2\theta}{T_\parallel}\right)\right], \tag{13}$$

where v is the absolute velocity, θ is the pitch angle of the velocity with respect to the quantization axis, and m and k are the electron mass and the Boltzmann constant, respectively. The function $f(v,\theta)$ is here normalized as

$$2\pi \iint f(v,\theta)v^2 \sin\theta \mathrm{d}v\mathrm{d}\theta = 1. \tag{14}$$

Figure 2 shows examples of $f(v,\theta)$ in the case of $T_\perp = 30\,\mathrm{eV}$ and $T_\parallel = 10\,\mathrm{eV}$ (a) and of $T_\perp = 30\,\mathrm{eV}$ and $T_\parallel = 100\,\mathrm{eV}$ (b). It is noted that when we focus our interest on a group of electrons having any constant absolute velocity, the number of electrons is larger in the higher temperature direction than in the lower temperature direction.

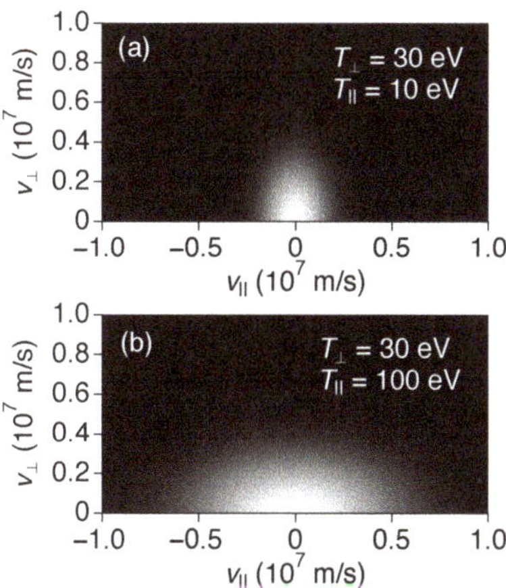

Figure 2. Examples of the electron velocity distribution function $f(v,\theta)$ for (**a**) $T_\perp = 30\,\mathrm{eV}$ and $T_\parallel = 10\,\mathrm{eV}$, and (**b**) $T_\perp = 30\,\mathrm{eV}$ and $T_\parallel = 100\,\mathrm{eV}$ cases.

The rate coefficients $C^{0,0}(1,p)$ and $C^{0,2}(1,p)$ are calculated [5] as

$$C^{0,0}(1,p) = \int Q_0^{0,0}(1,p) 4\pi f_0(v) v^3 \mathrm{d}v \tag{15}$$

and

$$C^{0,2}(1,p) = \int Q_0^{0,2}(1,p)[4\pi f_2(v)/5] v^3 \mathrm{d}v, \tag{16}$$

respectively, where $Q_0^{0,0}(1,p)$ and $Q_0^{0,2}(1,p)$ are the excitation and alignment creation cross sections, respectively, for the excitation from the ground state to the state p, and $f_0(v)$ and $f_2(v)$ are the coefficients of the expansion of $f(v,\theta)$ by the Legendre polynomials $P_K(\cos\theta)$ as

$$f(v,\theta) = \sum_K f_K(v) P_K(\cos\theta). \tag{17}$$

The coefficient $f_K(v)$ is explicitly given as

$$f_K(v) = \frac{2K+1}{2} \int f(v,\theta) P_K(\cos\theta) \sin\theta \, \mathrm{d}\theta. \tag{18}$$

The alignment creation cross section $Q_0^{0,2}(1,p)$ is derived from $Q_0^{0,0}(1,p)$ as [5].

$$Q_0^{0,2}(1,p) = (-1)^{J_p+J_s} \sqrt{\frac{2}{3}} (2J_p+1)^{-1} \begin{Bmatrix} J_p & J_p & 2 \\ 1 & 1 & J_s \end{Bmatrix}^{-1} A_{\mathrm{L}}(p,1) Q_0^{0,0}(1,p), \tag{19}$$

with A_L for the case when the excitation takes place with mono-energetic beam collisions. We adopt the cross section data by Bray [12] for $Q_0^{0,0}$ and by James [13] for P which is translated into A_L by Equation (11). It is noted that the data found in Refences [12,13] include both the $1\,^2S_{1/2}-2\,^2P_{1/2}$ and $1\,^2S_{1/2}-2\,^2P_{3/2}$ transitions. We here assume that 2/3 of the total cross section is for the $1\,^2S_{1/2}-2\,^2P_{3/2}$ transition, and the total polarization degree multiplied by 3/2 is for the $1\,^2S_{1/2}-2\,^2P_{3/2}$ transition (cf. Equation (5)). Figure 3 shows these elemental quantities relating to the $1\,^2S_{1/2}-2\,^2P_{3/2}$ transition. The opposite polarity between A_L and $Q_0^{0,2}$ is due to the 6-j symbol which is negative in the present case.

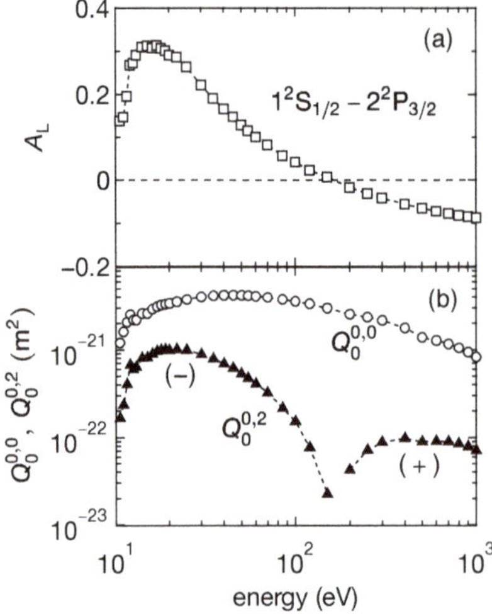

Figure 3. A_L values under an assumption of a mono-energetic beam collision experiment (**a**) and $Q_0^{0,0}$ and $Q_0^{0,2}$ (**b**) for the $1\,^2S_{1/2}-2\,^2P_{3/2}$ transition. The actual $Q_0^{0,2}$ labeled with (−) takes negative values.

The alignment destruction process is understood as the relaxation of the population imbalance among the magnetic sublevels. It is known that this process due to electron collisions has some correlation with the Stark broadening of the emission line from that state and its rate coefficient can be approximated by the half width of the Stark broadening [14]. Here, the Stark broadening data for the Lyman-α line by Stehlé [15] are adopted for evaluating $C^{2,2}(p,p)n_e$. The results are plotted in Figure 4. It is confirmed that the alignment destruction rate increases almost linearly with n_e.

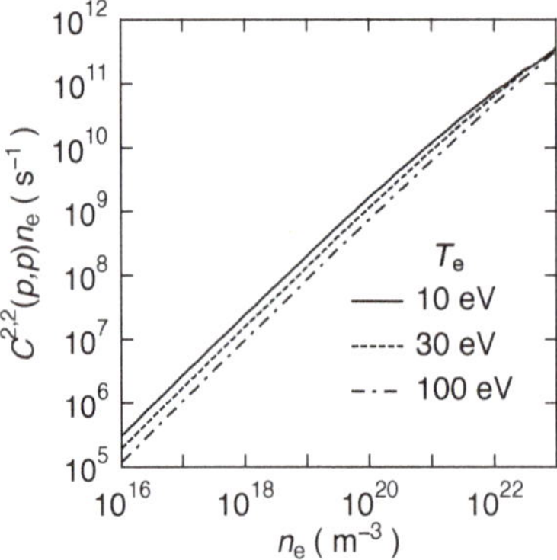

Figure 4. Alignment destruction rate $C^{2,2}(p,p)n_e$ evaluated from the Stark broadening width [15].

3. Results and Discussion

We have calculated A_L for the Lyman-α line for typical plasma conditions in LHD as an example. The quantization axis is taken in the direction of the magnetic field. In our previous experimental study, we found that the linearly polarized light intensity takes on a maximum (minimum) value in the direction perpendicular (parallel) to the magnetic field [9]. Therefore, it is natural to regard the magnetic field direction as the symmetry axis of the system. The calculation is made with $T_\perp$ fixed at 32 eV, where the experimental data analyzed later are borne in mind, while $T_\parallel$ is scanned in a range around the fixed $T_\perp$. We adopt several n_e values which cover a typical n_e range for the edge region of a magnetic fusion plasma. The polarization degree P for the Lyman-α line is derived from the A_L values with Equation (11), and the results are plotted in Figure 5.

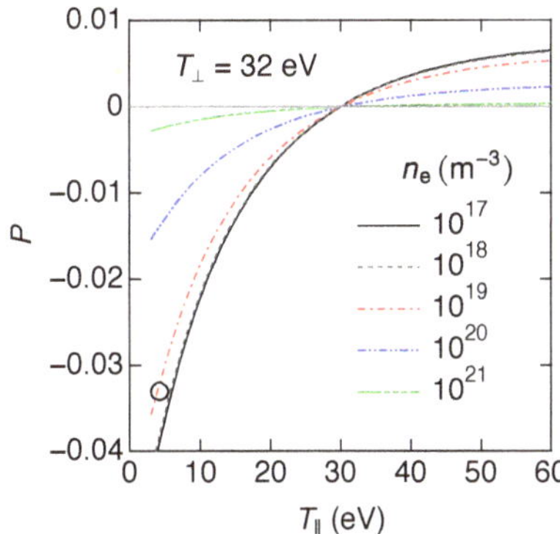

Figure 5. Example of the calculation results for P with several n_e values. $T_\perp$ is fixed at 32 eV and $T_\parallel$ is scanned. The open circle represents the combination of $n_e = 9.6 \times 10^{19}\ \mathrm{m}^{-3}$ and $P = -0.033$ corresponding to the experimental value in Ref. [9], from which $T_\parallel = 4.2\ \mathrm{eV}$ is derived.

It is confirmed that the line is unpolarized when $T_\parallel = T_\perp$, and the absolute polarization degree decreases with increasing n_e, which is caused by collisional averaging over the magnetic sublevels. The positive P values for $T_\parallel > T_\perp$ indicate that the π-light intensity is larger than that of σ-light, and negative P values for $T_\parallel < T_\perp$ mean the opposite condition. These results are inferred from the tendency of the elemental P data in Figure 3 as follows.

In the present T_e range at around 30 eV, the collision energy lower than 100 eV is dominant, and such collisions give rise to a positive A_L or P as seen in Figure 3a. A positive A_L or P means a higher intensity of linearly polarized light in the direction parallel with the electron beam axis than in the perpendicular direction (cf. Equation (1)). The condition of $T_\perp > T_\parallel$ means more electrons in the perpendicular direction than in the direction parallel to the magnetic field as seen in Figure 2, which causes a higher intensity in the perpendicular direction than in the direction parallel to the magnetic field. This tendency finally results in a negative A_L or P. The negative P values in the case of $T_\parallel < T_\perp$ can be understood similarly.

We have recently reported a value of $P = -0.033$ as an example in the actual measurement in LHD where the quantization axis is taken in the magnetic field direction [9]. In LHD, it is known that the radial location of neutral hydrogen emissions is almost fixed irrespective of the plasma condition [16–18], and local T_e and n_e at the emission location are obtained from the radial T_e and n_e profiles measured by the Thomson scattering diagnostic system. In the experiment where $P = -0.033$ is obtained, the T_e and n_e at the Lyman-α emission location are found to be 32 eV and $9.6 \times 10^{18}\ \mathrm{m}^{-3}$, respectively. Because the Thomson scattering diagnostic system for LHD measures light scattered by electrons moving predominantly in the direction perpendicular to the magnetic field, this T_e value can be regarded as $T_\perp$ in our model calculation. We now have $T_\perp$ and n_e at the emission location as well as P. Figure 5 indicates that $T_{\parallel}$ can be determined when $T_{\perp}$ and n_e are known so as to give the measured P. In the present case, $T_{\parallel} = 4.2\,\text{eV}$ is derived.

The atomic model developed in Section 2 assumes a simple mechanism for the formation of the excited level population. Here, the validity of the present model is examined with a collisional-radiative model (CR-model) for atomic hydrogen [19] which treats energy levels resolved only by the principal quantum number p.

The CR-model solves coupled rate equations for all of the excited levels considered in the model under the quasi-steady-state condition [19] for determining the population distribution over the excited levels. Because the present plasma is in the ionizing state [20], each excited level population is expressed as

$$n(p) = R_1(p) n_e n(1), \tag{20}$$

where $R_1(p)$ is called the population coefficient of the level p and is a function of n_e and T_e, and $n(1)$ stands for the ground state density. The CR-model derives $R_1(p)$ for all the excited levels considered.

By using the results of the CR-model, we have evaluated breakdowns of the population flows from and to the $p = 2$ level. Figure 6a shows the n_e dependence of component fractions of the inflow to the $p = 2$ level from other levels at $T_e = 10\,\text{eV}$. It is found that more than 90% of the inflow is dominated by the electron impact excitation, and the cascades from higher levels account for the remaining part of the inflow in the n_e range of our interest, i.e., from $10^{18}\,\text{m}^{-3}$ to $10^{19}\,\text{m}^{-3}$. We have confirmed that the results are hardly changed at $T_e = 30\,\text{eV}$. This result justifies the assumption regarding the populating process of the $p = 2$ level in the present model in Section 2.

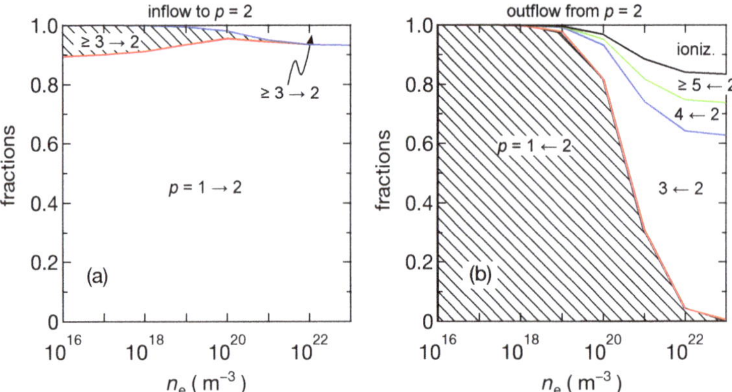

Figure 6. Fractions of breakdowns for the (**a**) population inflow to the level $p = 2$ and (**b**) population outflow from the level $p = 2$ as a function of n_e at $T_e = 10\,\text{eV}$. The numbers represent the principal quantum number of the levels and the arrows indicate the transition direction. The hatched and open areas indicate the radiative and collisional transitions, respectively. The label "ioniz". means the ionization.

Similar results concerning the outflow from the $p = 2$ are shown in Figure 6b. It is found that the radiative decay to the ground state predominates over other processes in the n_e range of our interest, which supports the assumption for the depopulating process from the $p = 2$ level in the model. Because collisional transition rates between different l-levels and j-levels are generally small as compared to those between p-levels under the conditions assumed here [21,22], the present model is regarded as a good approximation for the plasma considered.

In this paper, we have developed an atomic model for analyzing polarization states of the Lyman-α line where observations of the magnetically confined fusion plasma are borne in mind. The simplified model adopted is confirmed to be adequate through analyses for the populating mechanism of the $p = 2$ level which is the upper state of the Lyman-α line emission. An analysis of experimental data has been attempted with the present model, and the anisotropy is derived in terms of the difference in T_e between the magnetic field direction and the direction perpendicular to it. It is finally noted that the model developed in this paper is dedicated to the Lyman-α line, and therefore the anisotropy can be diagnosed in the limited narrow region where the Lyman-α line emission mainly takes place. However, the methodology can be easily transferred to other emission lines of other atoms and ions, so that an appropriate emission line can be chosen depending on the plasma region where the interest is focused.

Author Contributions: Conceptualization, methodology, and writing, M.G.; formal analysis, N.R.; investigation, M.G. and N.R. All authors have read and agreed to the published version of the manuscript.

Funding: This research was partly supported by JSPS KAKENHI Grant Number 18K03588 and by the National Institute for Fusion Science grant administrative budgets (ULHH028).

Acknowledgments: We thank Takashi Fujimoto for his helpful suggestions concerning the collisional process of polarization relaxation.

Conflicts of Interest: The authors declare no conflict of interest. The funders had no role in the design of the study; in the collection, analyses, or interpretation of data; in the writing of the manuscript, or in the decision to publish the results.

references

1. Qu, Z.S.; Fitzgerald, M; Hole, M.J. Analysing the impact of anisotropy pressure on tokamak equilibria. *Plasma Phys. Contr. Fusion* **2014**, *56*, 075007. [CrossRef]
2. Hole, M. J.; Qu, Z.; Pinches, S.; Schneider, M.; Arbina, I.L.; Mantsinen, M.J.; Sauter, O. The impact of anisotropy on ITER scenarios. *Nucl. Fusion* **2020**, *60*, 112010. [CrossRef]
3. Makino, T. Local and Fast Density Pump-out by ECRH in the LHD. *Plasma Fusion Res.* **2013**, *8*, 2402115. [CrossRef]
4. Rozhansky, V. Drifts, Currents, and Radial Electric Field in the Edge Plasma with Impact on Pedestal, Divertor Asymmetry and RMP Consequences. *Contrib. Plasma Phys.* **2014**, *54*, 508–-516. [CrossRef]
5. Fujimoto, T. Population-Alignment Collisional-Radiative Model. In *Plasma Polarization Spectroscopy*; Fujimoto, T., Iwamae, A., Eds.; Springer: Berlin, Germany, 2008; pp. 51–68.
6. Fujimoto, T.; Sahara, H.; Kawachi, T.; Kallstenius, T.; Goto, M.; Kawase, H.; Terumichi, Y. Polarization of impurity emission lines from a tokamak plasma. *Phys. Rev. E* **1996**, *54*, R2240–R2243. [CrossRef]
7. Herzog, O.; Weinheimer, J.; Rosmej, F.B.; Kunze, H.J.; Bertschinger, G.; Bitter, M.; Urnov, A. The Bragg-polarimeter at TEXTOR 94. In Proceedings of the Japan-US Workshop on Plasma Polarization Spectroscopy and The International Seminar on Plasma Polarization Spectroscopy, NIFS-PROC-37, Kyoto, Japan, 26–28 January 1998; pp. 33–38.
8. Kano, R.; Bueno, J.T.; Winebarger, A.; Auchère, F.; Narukage, N.; Ishikawa, R.; Carlsson, M. Discovery of Scattering Polarization in the Hydrogen Lyα Line of the Solar Disk Radiation. *ApJL* **2017**, *839*, L10. [CrossRef]
9. Ramaiya, N.; Goto, M.; Seguineaud, G.; Oishi, T.; Morita, S Measurement of polarization in Lyman-α line caused by anisotropic electron collisions in LHD plasma. *J. Quant. Spectrosc. Radiat. Transf.* **2020**, *260*, 107430. [CrossRef]
10. Iwamae, A.; Sato, T.; Horimoto, Y.; Inoue, K.; Fujimoto, T.; Uchida, M.; Maekawa, T. Anisotropic electron velocity distribution in an ECR helium plasma as determined from polarization of emission lines. *Plasma Phys. Contr. Fusion* **2005**, *47*, L41. [CrossRef]
11. Blum K. *Density Matrix Theory and Applications*, 2nd ed.; Plenum: New York, NY, USA, 1996.
12. Bray, I.; Stelbovics, A.T. Calculation of Electron Scattering on Hydrogenic Targets. *Adv. Atom. Mol. Opt. Phys.* **1995**, *35*, 209–254.
13. James, G.; Slevin, J.A.; Dziczek, D.; McConkey, J.W.; Bray, I. Polarization of Lyman-α radiation from atomic hydrogen excited by electron impact from near threshold to 1800 eV. *Phys. Rev. A Atom. Mol. Opt. Phys.* **1998**, *57*, 1787–1797. [CrossRef]
14. Hirabayashi, A.; Nambu, Y.; Hasuo, M.; Fujimoto, T. Disalignment of excited neon atoms due to electron and ion collisions. *Phys. Rev. A Atom. Mol. Opt. Phys.* **1988**, *37*, 83–88 [CrossRef] [PubMed]
15. Stehlé, C.; Hutcheon, R. Extensive tabulations of Stark broadened hydrogen line profiles. *Astron. Astrophys. Suppl. Ser.* **1999**, *140*, 93–97. [CrossRef]
16. Iwamae, A.; Hayakawa, M.; Atake, M.; Fujimoto, T.; Goto, M.; Morita, S. Polarization resolved H_α spectra from the large helical device: Emission location, temperature, and inward flux of neutral hydrogen. *Phys. Plasmas* **2005**, *12*, 042501. [CrossRef]
17. Iwamae, A.; Sakaue, A.; Neshi, N.; Yanagibayashi, J.; Hasuo, M.; Goto, M.; Morita, S. Hydrogen emission location, temperature and inward velocity in the peripheral helical plasma as observed with plasma polarization spectroscopy. *J. Phys. B: Atom. Mol. Opt. Phys.* **2010**, *43*, 144019. [CrossRef]
18. Goto, M.; Sawada, K. Determination of electron temperature and density at plasma edge in the Large Helical Device with opacity-incorporated helium collisional-radiative model. *J. Quant. Spectrosc. Radiat. Transf.* **2014**, *137*, 23–28. [CrossRef]

19. Sawada, K.; Fujimoto, T. Temporal relaxation of excited-level populations of atoms and ions in a plasma: Validity range of the quasi-steady-state solution of coupled rate equations. *Phys. Rev. E* **1994**, *49*, 5565–5573. [CrossRef] [PubMed]
20. Fujimoto, T. *Plasma Spectroscopy*; Oxford University Press: Oxford, UK, 2004.
21. Hutcheon, R.J.; McWhirter, R.W.P. The intensities of the resonance lines of highly ionized hydrogen-like ions. *J. Phys. B Atom. Mol. Optic. Phys.* **1973**, *6*, 2668–2683. [CrossRef]
22. Zygelman, B.; Dalgarno, A. Impact excitation of the n = 2 fine-structure levels in hydrogenlike ions by protons and electrons. *Phys. Rev. A* **1987**, *35*, 40850–4100. [CrossRef]

 MDPI

Article

Data for Beryllium–Hydrogen Charge Exchange in One and Two Centres Models, Relevant for Tokamak Plasmas

Petr A. Sdvizhenskii [1], Inga Yu. Tolstikhina [2], Valery S. Lisitsa [1,3], Alexander B. Kukushkin [1,3,*] and Sergei N. Tugarinov [4]

1 National Research Center "Kurchatov Institute", 123182 Moscow, Russia; Sdvizgenskii_PA@nrcki.ru (P.A.S.); Lisitsa_VS@nrcki.ru (V.S.L.)
2 P. N. Lebedev Physical Institute, 119991 Moscow, Russia; inga@lebedev.ru
3 National Research Nuclear University MEPhI (Moscow Engineering Physics Institute), 115409 Moscow, Russia
4 Institution "Project Center ITER", 123182 Moscow, Russia; S.Tugarinov@iterrf.ru
* Correspondence: Kukushkin_AB@nrcki.ru

Abstract: Data on the cross section and kinetic rate of charge exchange (CX) between the bare beryllium nucleus, the ion Be(+4) and the neutral hydrogen atom are of great interest for visible-range high-resolution spectroscopy in the ITER tokamak because beryllium is intended as the material for the first wall in the main chamber. Here an analysis of available data is presented, and the data needs are formulated. Besides the active probe signal produced by the CX of the diagnostic hydrogen neutral beam with impurity ions in plasma, a passive signal produced by the CX of impurity ions with cold edge plasma is also important, as it shows in observation data from the JET (Joint European Torus) tokamak with an ITER-like beryllium wall. Data in the range of a few eV/amu to ~100 eV/amu (amu stands for the atomic mass unit) needed for simulations of level populations for principal and orbital quantum numbers in the emitting beryllium ions Be(+3) can be obtained with the help of two-dimensional kinetic codes. The lack of literature data, especially for data resolved in orbital quantum numbers, has instigated us to make numerical calculations with the ARSENY code. A comparison of the results obtained for the one-centre Coulomb problem using an analytic approach and for the two-centre problem using numerical simulations is presented.

Keywords: charge exchange; cross section; tokamak plasmas; spectroscopy

Citation: Sdvizhenskii, P.A.; Tolstikhina, I.Y.; Lisitsa, V.S.; Kukushkin, A.B.; Tugarinov, S.N. Data for Beryllium –Hydrogen Charge Exchange in One and Two Centres Models, Relevant for Tokamak Plasmas. *Symmetry* **2021**, 13, 16. https://dx.doi.org/10.3390/sym13010016

Received: 8 December 2020
Accepted: 20 December 2020
Published: 23 December 2020

1. Introduction

The use of beryllium as a material for the first wall in the main chamber of the ITER tokamak requires detailed data on the cross sections of elementary atomic processes involving beryllium. Although such processes have been studied in detail, there are still types of processes for which the databases are not complete enough in the parameter ranges of interest. Such a process is the charge exchange of bare beryllium ions with hydrogen ions and its isotopes, which plays an important role in optical diagnostics of plasma in the visible spectral range. The charge exchange process can be written in the following general form:

$$A^{z+} + B^0(n_B) \to A^{(z-1)+}(n,l) + B^+ \tag{1}$$

It plays an important role for the Charge eXchange Recombination Spectroscopy (CXRS) diagnostics that Russia will supply to the ITER [1], as well as for other diagnostics of the Active Beam Spectroscopy type as a process that strongly affects the background signal (for more information, see, e.g., [2]). For such diagnostics, a diagnostic beam of neutral atoms is injected into the plasma. The plasma ion A^{z+} interacts with a neutral atom B^0 in a state with the principal quantum number n_B from the diagnostic beam and captures an electron from it. Usually, an electron is captured in an excited state $A^{(z-1)+}(n,l)$, where n, l are the principal and orbital quantum numbers, respectively. The excitation relaxes by the emission of radiation, which is collected by the optical system and delivered to the spectrometers.

The CXRS diagnostics measures important plasma parameters, such as impurity concentration and distribution, ion temperature, and plasma rotation velocity profiles. For reliable interpretation of measurements, predictive modelling of the plasma emission spectra is necessary, which, in particular, requires information on the cross sections of the charge exchange reaction (1).

Besides the active signal produced by the CX of diagnostic neutral hydrogen beam with impurity ions in the plasma, the passive signal produced by the CX of impurity ions with atoms in the cold edge plasma is also important, as follows from the available observational data from the JET (Joint European Torus) tokamak with ITER-like beryllium wall (see, e.g., comments and references in [2]).

This paper is focused on data on the cross section of charge exchange of bare beryllium nuclei with neutral atoms of hydrogen isotopes:

$$Be^{4+} + H^{0}(n_{H}) \rightarrow Be^{3+}(n,l) + H^{+} \quad (2)$$

(Herein parentheses are the quantum numbers of the state of an atom or ion). We present (i) an overview of the data available in the literature and databases, and (ii) new data recently calculated using the ARSENY code [3,4], in the range from a few eV/amu to ~1000 eV/amu (amu stands for the atomic mass unit) especially the data resolved in orbital quantum numbers, which are needed for modelling the level populations of emitting beryllium ions Be(+3) with the help of two-dimensional, in principal and orbital quantum numbers, kinetic codes.

The following transitions of the hydrogen-like beryllium ion Be IV between the states with certain principal quantum numbers will be used to implement CXRS diagnostics on the ITER: 4658.42 Å (6–5 transition) and 4685.4 Å (8–6 transition). Therefore, it is important to have data for charge exchange cross sections at high levels that contribute to the spectral transitions used for measurements.

A comparison of the results obtained for the one-centre Coulomb problem using the analytical approach [5] and the two-centre problem using numerical simulations [6–16] is presented.

2. Materials and Methods

One and Two Centres Symmetry Problems in CXRS

The problem of charge exchange recombination of a neutral hydrogen atom on a nucleus with a charge Z, which is a classic example of the evolution of atomic states between two Coulomb centres, was considered within the framework of the one-centre model. The charge exchange process here is due to the intersection of quasi-molecular terms in the system of repulsive centres. The transition probability was determined by the exchange matrix element at the intersection points and was calculated by the Landau-Zener method (see [17]). The further evolution of the recharged states between two points of convergence and scattering of nuclei occurred under the action of an electric field created by a recharged nucleus moving along a classical trajectory. This field, varying in magnitude and direction, caused a mixing of the initially degenerate one-centre states with respect to orbital angular momenta (the effect of rotation of the internuclear axis). Thus, the initially populated state with a principal quantum number n of the order of $Z^{2/3}$ due to charge exchange from the ground state of the hydrogen atom, which possessed zero orbital angular momentum, turned out to be strongly mixed over states with other orbital angular momenta due to the effects of the rotation of the internuclear axis. These effects led to the population of states with predominantly large orbital angular momenta. This phenomenon was observed already in the first experiments on CXRS [18] of carbon nuclei, where the increase in the luminescence at the transition from the initially populated state $n = 4$ to the nearest level $n = 3$ did not occur, while the intensity of the 3–2 radiative transition increased noticeably. This is evidently explained by the selection rules for the radiation cascade between excited states with the maximum orbital momenta. Of course, transitions between the levels $n = 4$ and $n = 2$ are possible for different values of the orbital angular momentum due to the mixing of states with different orbital angular momenta in collisions with plasma ions. However, the predominance of radiative transitions with a decrease in the orbital angular momentum over similar transitions with an increase in it correlates well with the observed effect.

The one-centre model is attractive due to the ability to follow the above effects analytically. This possibility is due to the symmetry of the Coulomb field in the one-centre problem. This symmetry makes it possible to construct the exact evolution of atomic states under the action of the ZR^{-2} Coulomb force acting on the degenerate one-centre states of the target atom. This force has the same decay law with distance R as the centrifugal energy of an electron. This makes it possible to construct "dynamic" terms of the system evolving under the action of the indicated forces, parametrically depending on the relative velocity of the nuclei [5].

Thus, we can speak of the charge exchange transition directly into these dynamic terms, which directly take into account the effects of the rotation of the internuclear axis.

However, further analysis showed that the charge exchange process, especially at low energies, is extremely sensitive to the structure of electronic terms, and the one-centre approximation becomes insufficient. In this regard, numerous calculations have been performed on the two-centre basis of wave functions. These calculations use a more general symmetry of the two-centre Coulomb problem associated with the separation of variables in hyperspherical coordinates. In view of the complex nature of the dependence of the terms on the internuclear distance, the corresponding calculations are carried out by numerical methods.

In this work, we present the results of such calculations for the charge exchange of atomic hydrogen in the ground and first excited states on the nuclei of beryllium, which is the main impurity element of the planned thermonuclear tokamak reactor. At the same time, a comparison is made both with the results of calculations in the one-centre model and with two-centre models among themselves. Such a comparison allows one to judge the limits of applicability of models based on the use of one- and two-centre symmetries of the Coulomb field.

For the calculation of the charge exchange cross sections for slow collisions, the ARSENY code, based on the method of hidden crossings, is used [3]. In the adiabatic approximation, radial inelastic transitions occur in the regions of the closest approach of potential curves and are decomposed into a sequence of individual two-level transitions via hidden crossings. Electronic energies are the eigenvalues of the two-centre Coulomb problem [4], which is separable in the prolate spheroidal coordinates and is solved for the complex internuclear distance R.

In the adiabatic theory, the charge exchange transitions occur at the internuclear distances where the electronic wave function changes rapidly. This happens when the non-adiabatic coupling has its maximum. Transitions caused by the radial coupling take place in the hidden crossings (branch points) at complex internuclear separation R, where the electronic energies of two states are equal. Hidden crossings arise when the full-dimensional classical trajectory of the electron collapses into an unstable periodic orbit. They are invisible on the plot of the adiabatic potential curves at the real value of the adiabatic parameter R and require direct calculation in the complex R-plane.

Rotational coupling, associated with the rotation of the internuclear axis in close collisions, induces transitions between electronic states, which are degenerate in the limit of the united atom. The potential curves of these states have an exact crossing at complex R values ($\mathrm{Re}(R) = 0$). Since the scattering angle depends on the reduced mass, the trajectories of the heavy particles in the reactions with H, D and T are different. This, in turn, leads to a difference in the corresponding cross sections.

3. Results

3.1. A Brief Review of the Data Available in OPEN-ADAS

The figures in this section show the data for the cross sections for reaction (2), taken from the OPEN-ADAS database [6,7], and their comparison with cross sections from various sources, depending on the relative velocity v and the collision energy E, expressed in units of eV/amu (amu stands for the atomic mass unit). Collision energy $E = 2.5$ eV/amu corresponds to the relative velocity of colliding particles $v \approx 0.01$ a.u. (1 a.u. $\approx 2.188\ 10^6$ m/s); collision energy $E = 1$ keV/amu corresponds to $v \approx 0.2$ a.u.

Figure 1 shows a comparison of the data on the charge exchange cross sections for hydrogen in the ground state, taken from the OPEN-ADAS database [6,7], paper [8], with calculations within the one-centre model [5].

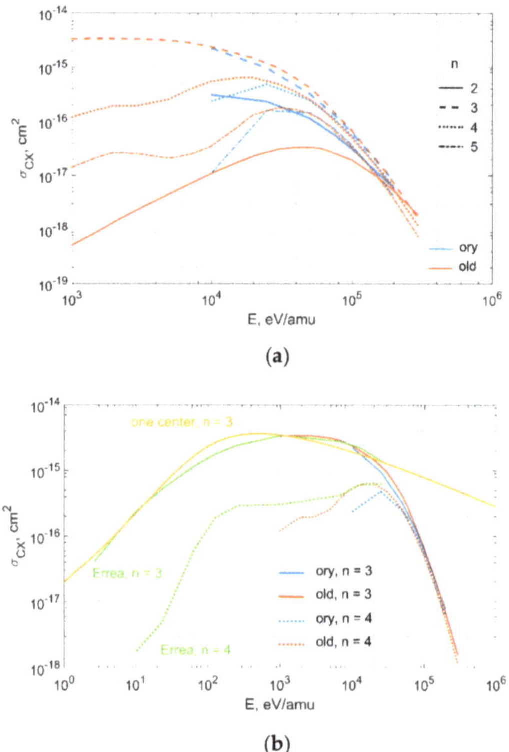

Figure 1. The cross sections [6–8] for reaction (2), for the values of principal quantum number n indicated in the legend: (**a**) the data from the OPEN-ADAS database [6,7]; (**b**) the comparison of data from the OPEN-ADAS database [6,7] with the results from [8] and with calculations within the one-centre model [5].

The curve names ("ory" and "old") in the legend in Figure 1 refer to the filename in OPEN-ADAS [6,7] from which the data were taken. The files in OPEN-ADAS have names such as qcx#h0_*#be4.dat, where instead of * are the words specified in the legend. Thus, if the curve is labeled "ory" in the legend, it means that it is plotted from the data from the qcx#h0_ory#be4.dat file. The green curves (solid for the population of the atomic level with principal quantum number $n = 3$ and dotted for the population of level with $n = 4$) show the data from Errea 1998 [8], where the results for cross sections at low collision energies are available. The yellow curve shows the result of calculations within the one-centre model [5] for populating at the level of $n = 3$.

From Figure 1, one can see that the data available in OPEN-ADAS cover the region of medium and high collision energies, while data for low energies are not available.

Figure 2 presents a comparison of the data from the OPEN-ADAS database for charge exchange cross sections for hydrogen in the ground and excited states.

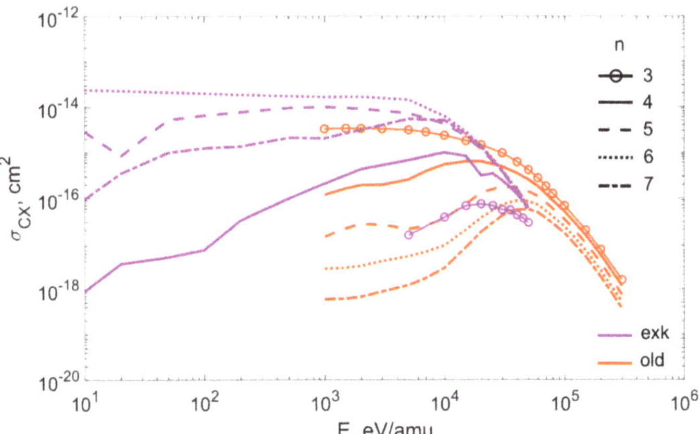

Figure 2. Comparison of the cross sections taken from [6,7] for reaction (2) for the principal quantum number of the state in the hydrogen atom $n_H = 1$ (brown curves) and $n_H = 2$ (violet curves), i.e., in the latter case, charge exchange occurs with hydrogen in the excited state. Other notations are the same as in Figure 1.

Figure 2 shows that the charge exchange cross sections for hydrogen in the excited state are higher than those for hydrogen in the ground state. In the first case, the largest cross section corresponds to the charge exchange with the population of level with the principal quantum number $n = 6$, in the second—the level with $n = 3$.

Figures 3–5 show the data from the qcx#h0_en2_kvi#be4.dat file from OPEN-ADAS for charge exchange from excited hydrogen with a population of levels with different n and l. In Figure 3, one can see a comparison of these data with cross sections from several other sources, while Figures 4 and 5 show a detailed l-resolved comparison with calculations with the ARSENY code.

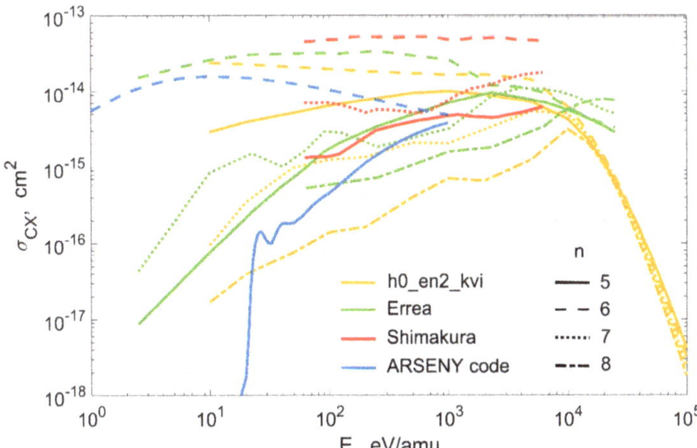

Figure 3. The charge exchange cross sections for reaction (2) with excited hydrogen for the population of levels with different n, according to the data from the qcx#h0_en2_kvi#be4.dat file [6,7] (yellow curves), [8] (green curves), [9] (red curves) and calculations using the ARSENY code (blue curves).

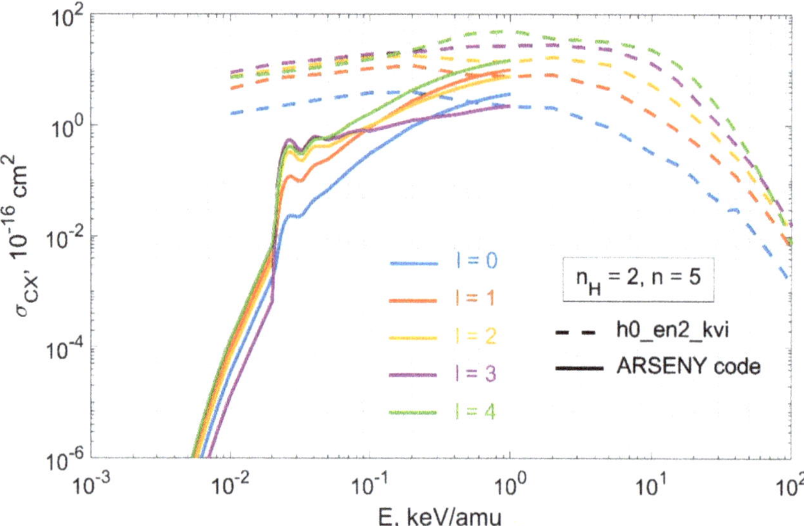

Figure 4. The charge exchange cross sections for reaction (2) with excited hydrogen for electron capture to the state with principal quantum number $n = 5$ and orbital quantum number l. Solid curves—calculations using the ARSENY code, dashed curves—data from qcx#h0_en2_kvi#be4.dat file [6,7].

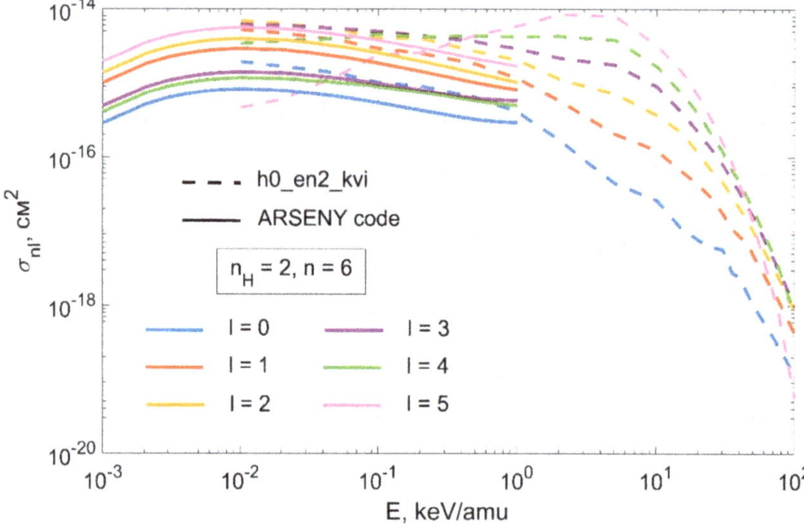

Figure 5. The charge exchange cross sections for reaction (2) with excited hydrogen, according to the data from the qcx#h0_en2_kvi#be4.dat file [6,7] and calculations using the ARSENY code.

3.2. Comparison of Cross Sections from Special Issue of Physical Scripta

Figure 6, Figure 7, Figure 8, Figure 9 below show the comparison of charge exchange cross sections for reaction (2), taken from [10] and [11]. Figure 10 shows the dependence of charge exchange cross sections for reaction (2) on the orbital quantum number l for different values of the principal quantum number n.

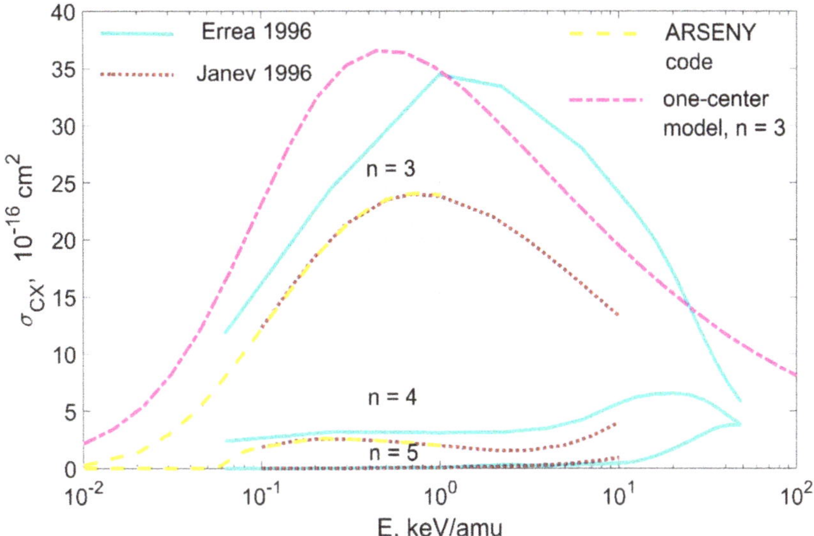

Figure 6. Charge exchange total cross sections for reaction (2) for electron capture to the state with the principal quantum number n (the values of n are shown in the figure). Solid curves—data from [10], dotted curves—data from [11], dashed curves—calculations using the ARSENY code, dash-dotted curve—calculations within the one-centre model [5], for the population of level with $n = 3$.

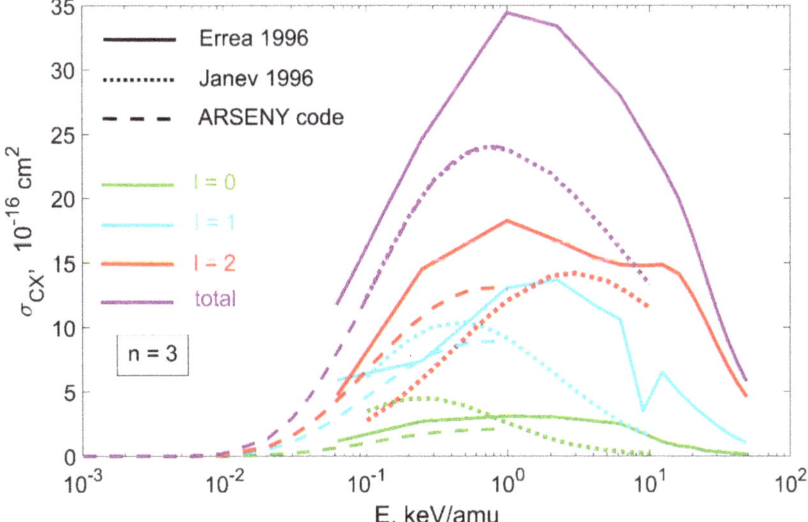

Figure 7. Partial charge exchange cross sections for reaction (2) for electron capture to the state with the principal quantum number $n = 3$ and orbital quantum number l (green curves—$l = 0$, blue curves—$l = 1$, red curves—$l = 0$, violet curves—total). Solid curves—data from [10], dotted curves—data from [11], dashed curves—calculations with the ARSENY code.

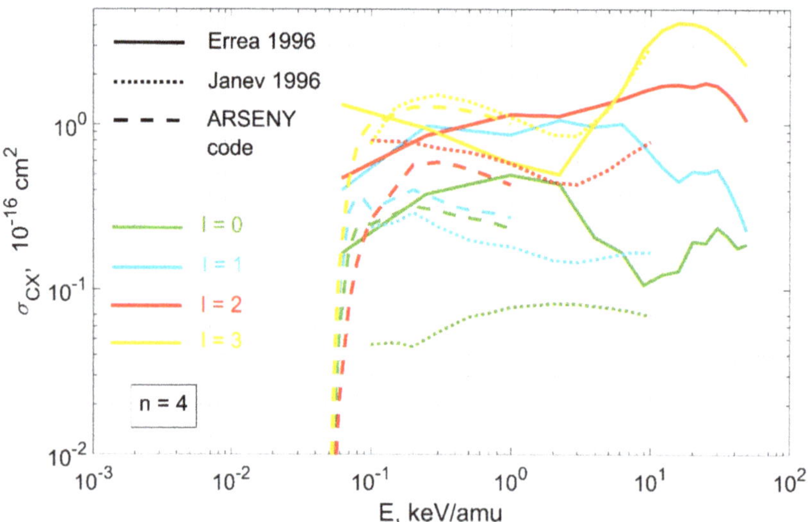

Figure 8. The same as in Figure 7, but for $n = 4$; yellow curves—$l = 3$.

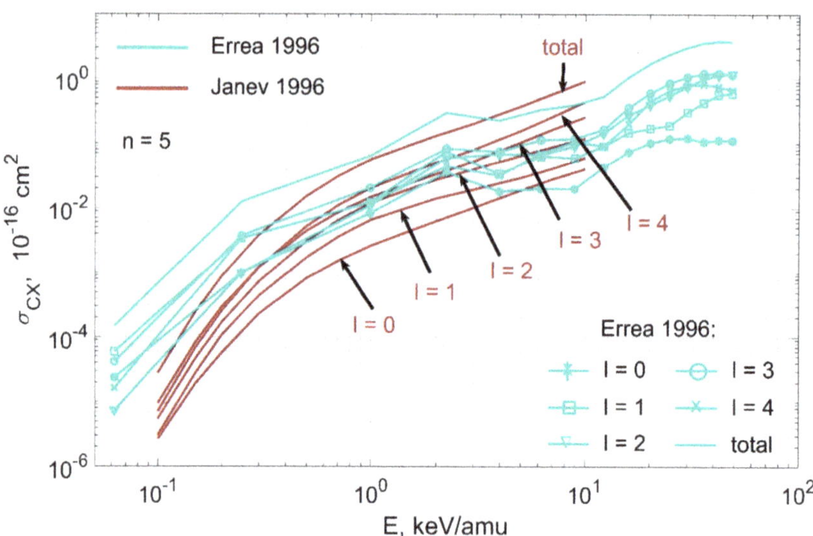

Figure 9. Same as in Figure 7, but for $n = 5$.

3.3. A Survey of Data on Cross Sections from Various Sources

It can be seen from the Figures in previous subsections that in the range of collision energies below 100 eV/amu, there is practically no data resolved in orbital quantum numbers l.

Figures 11 and 12 show a comparison of the charge exchange cross sections calculated by various methods for reaction (2), taken from various sources.

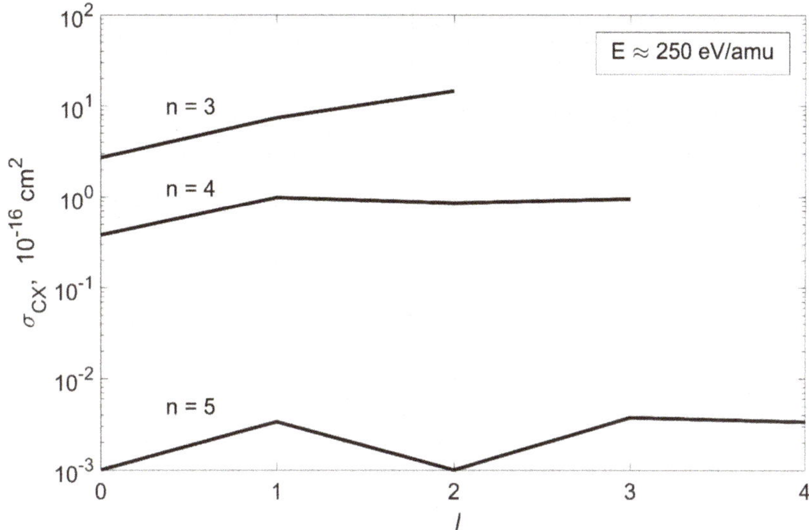

Figure 10. The dependence of charge exchange cross section for reaction (2) on the orbital quantum number l for different values of n (the values of n are shown in the figure). The data from [10] are used.

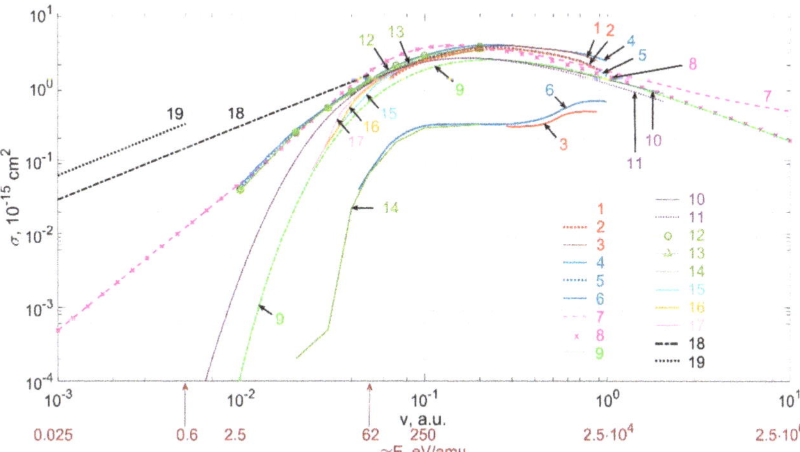

Figure 11. A survey of data on cross sections for reaction (2). The sources of data are indicated in the legend. The notations of curves are explained in the text below.

The notations of the curves in Figures 11 and 12 are as follows. Curves 1, 2 and 3 stand for the calculation [12] of the total cross section and partial cross sections for $n = 3$ and $n = 4$, respectively, using 21-state atomic-orbital expansion. Curves 4, 5 and 6 stand for the calculation [8] of the total cross section and partial cross sections for $n = 3$ and $n = 4$, respectively, using molecular expansion with semiclassical and quantal calculations for 96- and 17-state basis sets. Curves 7 and 8 stand for our calculations of the total cross section and partial cross section for $n = 3$, respectively, in the frame of the Landau-Zener model with rotation taken into account [5], while curve 9 is the same but without rotation taken into account [5]. Curves 10 and 11 stand for the total cross section and partial cross section for $n = 3$, respectively, according to the Kronos database [13]; the model is described in [14].

Curves 12, 13 and 14 stand for the calculation [15] of the total cross section and partial cross sections for $n = 3$ and $n = 4$, respectively, using the hyperspherical close-coupling (HSCC) approach. Curve 15 stands for the calculation [16] for hydrogen; curve 16 stands for the calculation [16] for deuterium $Be^{4+} + D^0(1s) \rightarrow Be^{3+}(n) + D^+$ (for more details on the isotopic effect see, e.g., [19] or review [20]); the curve 17 stands for the calculation [16] for tritium: $Be^{4+} + T^0(1s) \rightarrow Be^{3+}(n) + T^+$; calculations were performed in the framework of the adiabatic theory of transitions in slow collisions using the ARSENY code, based on the hidden crossings method. The curve 18 stands for low velocities asymptotic (2.18) in [5] with $| b_+(R)/b_-(R) | = 1$, $n = 3$, $Z = 4$, $R_0 = R_{n=3}$, where R_n is defined by (2.2) in [5]; V is defined by (1.1) in [5].

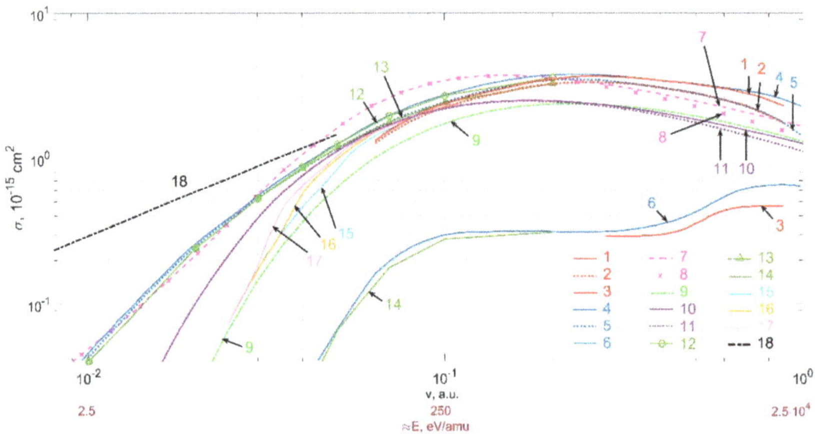

Figure 12. An enlarged fragment of Figure 11.

4. Discussion

In the lack of experimental data, special attention must be paid to the accuracy and reliability of calculations. The most reliable are the results obtained using the atomic/molecular orbital method or their variations. One of the drawbacks of calculations by the Landau-Zener model is the impossibility of calculating the cross sections for charge exchange with the population of levels with the principal quantum number $n \geq Z$, where Z is the ion charge.

Figures 1–5 in Section 3.1 show the data for the cross section for reaction (2) taken from the OPEN-ADAS database [6,7] and their comparison both with each other and with the cross sections from several other sources. Figure 1 also shows the calculations within the one-centre model [5], while Figures 4 and 5 show a detailed l-resolved comparison with calculations using the ARSENY code [3,4].

Section 3.2 is devoted to a comparison of the cross sections from [10,11], calculations in the framework of the one-centre model, and calculations using the ARSENY code are presented there as well.

Figures 11 and 12 in Section 3.3 show a survey of data for cross sections, taken from various sources with different models and approaches used for calculations.

It can be seen from Section 3 that in the literature and existing databases (e.g., OPEN-ADAS [6,7]), most of the available information on the charge exchange cross sections covers the range of energies of interest primarily for calculating the active signal of the Charge eXchange Recombination Spectroscopy (CXRS), i.e., for diagnostic beams with energies of tens and hundreds of keV. At the same time, for estimates of the passive charge exchange signal (see, e.g., the recently proposed algorithm [21]), when the energy of neutral atoms coming from the wall is in the range from few eV to tens or hundreds of eV, the data are not enough, especially for *l*-resolved data. Therefore, for the charge exchange reactions of interest, additional theoretical calculations of the cross sections may be necessary, which are needed for simulations of level populations of the emitting beryllium ions Be^{3+} using two-dimensional, in principal and orbital quantum numbers, kinetic codes, such as the *nl*-KINRYD code [22].

5. Conclusions

An analysis of the available data on the cross sections of the charge exchange reaction of bare beryllium nuclei on hydrogen isotopes was carried out. The dependence of the cross sections on the relative nuclear energy, which are selective both in the principle and orbital momentum quantum numbers, were presented in a wide range of relative energies that are relevant for the CXRS diagnostics of the beryllium impurity, the main component of the plasma facing materials in the thermonuclear tokamak-reactor ITER. The charge exchange cross sections from the first excited energy states of hydrogen were presented as well. The approaches based on one- and two-centres Coulomb symmetry were discussed in detail. A comparison was made of a number of modern approaches to simulations of charge exchange cross sections. It was shown that there was a lack of data in the region of low (below 100 eV/amu) collision energies for beryllium, which is now reduced due to calculations using the ARSENY code. However, there is still a lack of data in the entire space of states (populations of other states in beryllium, and other impurities, such as carbon), which is of interest for modelling the passive charge exchange signal of the CXRS diagnostic of impurities in the ITER tokamak and other thermonuclear fusion facilities.

Author Contributions: Conceptualisation, V.S.L. and A.B.K.; methodology, I.Y.T., V.S.L. and P.A.S.; software, I.Y.T. and P.A.S.; validation, P.A.S., I.Y.T. and V.S.L.; formal analysis, P.A.S.; resources, S.N.T.; writing—original draft preparation, P.A.S., I.Y.T. and V.S.L.; writing—review and editing, V.S.L. and A.B.K.; visualisation, P.A.S.; supervision, S.N.T. and A.B.K.; project administration, S.N.T. All authors have read and agreed to the published version of the manuscript.

Funding: This research was funded in part by the Institution "Project Center ITER".

Institutional Review Board Statement: Not applicable.

Informed Consent Statement: Not applicable.

Data Availability Statement: Not applicable.

Acknowledgments: The authors are grateful to M. G. O'Mullane for discussing the subject of this paper.

Conflicts of Interest: The authors declare no conflict of interest.

References

1. Tugarinov, S.N.; Beigman, I.L.; Vainshtein, L.A.; Dokuka, V.N.; Krasil'nikov, A.V.; Naumenko, N.N.; Tolstikhina, I.Y.; Khairutdinov, R.R. Development of the concept of charge-exchange recombination spectroscopy for ITER. *Plasma Phys. Rep.* **2004**, *30*, 128–135. [CrossRef]
2. Von Hellermann, M.; de Bock, M.; Marchuk, O.; Reiter, D.; Serov, S.; Walsh, M. Simulation of Spectra Code (SOS) for ITER Active Beam Spectroscopy. *Atoms* **2019**, *7*, 30. [CrossRef]
3. Solov'ev, E.A. *Workshop on Hidden Crossings in Ion-Collisions and in Other Nonadiabatic Transitions*; Harvard Smithonian Centre for Astrophysics: Cambridge, MA, USA, 1991.
4. Komarov, I.; Ponomarev, L.; Slavyanov, S. *Spheroidal and Coulomb Spheroidal Functions*; Nauka: Moscow, Russia, 1976.
5. Abramov, V.A.; Baryshnikov, F.F.; Lisitsa, V.S. Charge transfer between hydrogen atoms and the nuclei of multicharged ions with allowance for the degeneracy of the final states. *Sov. Phys. JETP* **1978**, *47*, 469–477. Available online: http://www.jetp.ac.ru/cgi-bin/dn/e_047_03_0469.pdf (accessed on 7 December 2020).
6. OPEN-ADAS (Atomic Data and Analysis Structure). Available online: http://open.adas.ac.uk/ (accessed on 7 December 2020).
7. Summers, H.P.; O'Mullane, M.G. Atomic data and modelling for fusion: The ADAS Project. *AIP Conf. Proc.* **2011**, *1344*, 179–187. [CrossRef]
8. Errea, L.F.; Harel, C.; Jouin, H.; Méndez, L.; Pons, B.; Riera, A. Quantal and semiclassical calculations of charge transfer cross sections in Be^{4+} + H collisions for impact energies of 2.5 eV amu^{-1} $< E <$ 25 keV amu^{-1}. *J. Phys. B At. Mol. Opt. Phys.* **1998**, *31*, 3527–3547. [CrossRef]

9. Shimakura, N.; Kobayashi, N.; Honma, M.; Nakano, T.; Kubo, H. Electron capture from excited hydrogen atoms by highly charged beryllium and carbon ions. *J. Phys. Conf. Ser.* **2009**, *163*, 012045. [CrossRef]
10. Errea, L.F.; Gorfinkiel, J.D.; Harel, C.; Jouin, H.; Macias, A.; Méndez, L.; Pons, B.; Riera, A. Total and partial cross-sections of electron transfer processes with hydrogen gas targets: Be^{4+}, B^{5+} + H(1s), H(2s). *Phys. Scr.* **1996**, *1996*, 27–33. [CrossRef]
11. Janev, R.K.; Solov'ev, E.A.; Ivanovski, G. State-selective electron capture in slow Be^{4+} + H(1s) collisions. Calculations by hidden crossing method. *Phys. Scr.* **1996**, *1996*, 43–48. [CrossRef]
12. Fritsch, W.; Lin, C.D. Atomic-orbital-expansion studies of electron transfer in bare-nucleus Z (Z= 2, 4–8)—Hydrogen-atom collisions. *Phys. Rev. A* **1984**, *29*, 3039. [CrossRef]
13. Kronos Database. Available online: https://www.physast.uga.edu/research/stancil-group/atomic-molecular-databases/kronos (accessed on 7 December 2020).
14. Mullen, P.D.; Cumbee, R.S.; Lyons, D.; Stancil, P.C. Charge Exchange-induced X-Ray Emission of Fe XXV and Fe XXVI via a Streamlined Model. *ApJS* **2016**, *224*, 31. [CrossRef]
15. Le, A.T.; Hesse, M.; Lee, T.G.; Lin, C.D. Hyperspherical close-coupling calculations for charge transfer cross sections in Si^{4+} + H(D) and Be^{4+} + H collisions at low energies. *J. Phys. B At. Mol. Opt. Phys.* **2003**, *36*, 3281–3293. [CrossRef]
16. Tolstikhina, I.Y.; Litsarev, M.S.; Kato, D.; Song, M.Y.; Yoon, J.S.; Shevelko, V.P. Collisions of Be, Fe, Mo and W atoms and ions with hydrogen isotopes: Electron capture and electron loss cross sections. *J. Phys. B At. Mol. Opt. Phys.* **2014**, *47*, 035206. [CrossRef]
17. Landau, L.; Lifshitz, E.M. *Quantum Mechanics: Non-Relativistic Theory*, 2nd ed.; Addison-Wesley: Reading, MA, USA, 1965.
18. Isler, R.C. Observation of the Reaction $H^0 + O^{8+} \rightarrow H^+(O^{7+})$ during Neutral-Beam Injection into ORMAK. *Phys. Rev. Lett.* **1977**, *38*, 1359–1362. [CrossRef]
19. Tolstikhina, I.Y.; Tolstikhin, O.I. Effect of electron-nuclei interaction on internuclear motions in slow ion-atom collisions. *Phys. Rev. A* **2015**, *92*, 042707. [CrossRef]
20. Tolstikhina, I.Y.; Shevelko, V.P. Influence of atomic processes on charge states and fractions of fast heavy ions passing through gaseous, solid, and plasma targets. *Phys.-Uspekhi* **2018**, *61*, 247–280. [CrossRef]
21. Sdvizhenskii, P.A.; Kukushkin, A.B.; Levashova, M.G.; Lisitsa, V.S.; Neverov, V.S.; Serov, S.V.; Tugarinov, S.N. Algorithm of calculating the passive signal of tokamak edge plasma for CXRS diagnostics. In Proceedings of the 46th EPS Conference on Plasma Physics, Milan, Italy, 8–12 July 2019; European Conference Abstracts. Volume 43C. Available online: http://ocs.ciemat.es/EPS2019ABS/pdf/P4.1006.pdf (accessed on 7 December 2020).
22. Kadomtsev, M.B.; Levashova, M.G.; Lisitsa, V.S. Semiclassical theory of the radiative-collisional cascade in a Rydberg atom. *J. Exp. Theor. Phys.* **2008**, *106*, 635–649. [CrossRef]

Article

Hypersonic Imaging and Emission Spectroscopy of Hydrogen and Cyanide Following Laser-Induced Optical Breakdown

Christian G. Parigger [1,2,*], Christopher M. Helstern [1,2] and Ghaneshwar Gautam [3]

1 Center for Laser Applications, University of Tennessee Space Institute, 411 B. H. Goethert Parkway, Tullahoma, TN 37388, USA; Chris.Helstern@gmail.com
2 Physics and Astronomy Department, University of Tennessee, Knoxville, TN 37996, USA
3 Fort Peck Community College, 605 Indian Avenue, Poplar, MT 59255, USA; ggautam@fpcc.edu
* Correspondence: cparigge@tennessee.edu; Tel.: +1-931-841-5690

Received: 25 November 2020; Accepted: 17 December 2020; Published: 19 December 2020

Abstract: This work communicates the connection of measured shadowgraphs from optically induced air breakdown with emission spectroscopy in selected gas mixtures. Laser-induced optical breakdown is generated using 850 and 170 mJ, 6 ns pulses at a wavelength of 1064 nm, the shadowgraphs are recorded using time-delayed 5 ns pulses at a wavelength of 532 nm and a digital camera, and emission spectra are recorded for typically a dozen of discrete time-delays from optical breakdown by employing an intensified charge-coupled device. The symmetry of the breakdown event can be viewed as close-to spherical symmetry for time-delays of several 100 ns. Spectroscopic analysis explores well-above hypersonic expansion dynamics using primarily the diatomic molecule cyanide and atomic hydrogen emission spectroscopy. Analysis of the air breakdown and selected gas breakdown events permits the use of Abel inversion for inference of the expanding species distribution. Typically, species are prevalent at higher density near the hypersonically expanding shockwave, measured by tracing cyanide and a specific carbon atomic line. Overall, recorded air breakdown shadowgraphs are indicative of laser-plasma expansion in selected gas mixtures, and optical spectroscopy delivers analytical insight into plasma expansion phenomena.

Keywords: laser–plasma interactions; plasma dynamics and flow; hypersonic flows; optical emission spectroscopy; hydrogen; cyanide; Abel inversion; astrophysics; white dwarf stars

1. Introduction

Laser-plasma research is experiencing remarkable interest in laser-induced optical breakdown (LIBS) [1], in part due to success in analytical chemistry, in a volley of engineering applications, or in dedicated diagnosis that may extend to technology-driven changes in the medical field. This work is concerned with experiments and analysis of phenomena associated with pulsed, nanosecond radiation: Optical breakdown is accomplished by focusing a laser beam to irradiances above threshold for local lightning or transient plasma generation in gaseous media. For plasma generation with focused nanosecond laser pulses, the initial portion of the laser pulse energy generates optical breakdown and the remaining portion interacts with the evolving plasma. Micro-plasma imaging is of general interest in the laser-induced breakdown experiments, this includes application of LIBS diagnostic in gases and near liquid or solid surfaces. The analysis and interpretation of observed expansion dynamics can be significantly alleviated when including symmetry considerations.

Optical radiation that is well-above optical breakdown threshold in air, hydrogen, and molar 1:1 CO_2:N_2 molar mixture at or near ambient laboratory conditions, causes multiple breakdown spots in focus. One may associate these spots with peaks in computational maps of the focal area, especially when considering a single lens and spherical aberration [2]. As one reduces the irradiance to the threshold values for the gases of interest, the number of separate breakdown spots diminishes down to one. However, there is always a desire to obtain more diagnostic signal, and reasonably repeatable signals that favor use of radiation that is perhaps of the order of one magnitude, or more, above threshold. This work aims to investigate occurrence of reasonable spherical symmetry in gas breakdown dynamics, and subsequently, to apply Abel inversion techniques for diagnostics of the expanding plasma. Of interest in this work is optical breakdown in air [3], early breakdown phenomena in hydrogen gas [4], and measurement of the diatomic molecule cyanide (CN) [5,6], including spatial distribution for time delays of the order of one microsecond form initiation of optical breakdown.

2. Experimental Arrangement and Methods

2.1. Shadowgraphs

Shadowgraphs of breakdown plasma are used to visualize plasma expansion [3]. It is imperative to capture shadowgraphs for plasma-excitation energies that were used for time-resolved spectroscopy. A Q-switched neodymium-doped yttrium aluminum garnet Nd:$Y_3Al_5O_{12}$ (Nd:YAG) laser (Quantel model Q-smart 850, USA) is operated at a fundamental wavelength of 1064 nm and a 10 Hz repetition rate to deliver full-width-at-half-maximum (FWHM) 6 ns laser radiation with an energy of 850 mJ per pulse. Beam splitters and apertures were used to attenuate laser-pulse energy from 850 to 170 mJ per pulse. Additionally, another Q-switched Nd:YAG laser (Continuum Surelite model SL I-10, USA) was frequency-doubled to operate at its second harmonic wavelength, 532 nm, with 5 ns pulses. Optical breakdown events generated with 850 mJ pulses in this work clearly show multiple breakdown spots early in the plasma expansion, however, use of 170 mJ pulses diminishes the number of breakdown spots, leading to an earlier appearance of near-spherical plasma expansion.

A 100 mm plano-convex lens, f/10 focusing, is used to focus the 1064 nm IR beam. The 532 nm green laser beam is reflected by two prisms and three mirrors to produce a shadowgraph of the plasma. A beam expander and reflections by the prisms and mirrors caused the beam to become wide enough to cover the entire plasma volume for shadowgraph imaging. Figure 1 shows a schematic of the experimental setup, and Figure 2 communicates a photograph of a typical shadowgraph experiment.

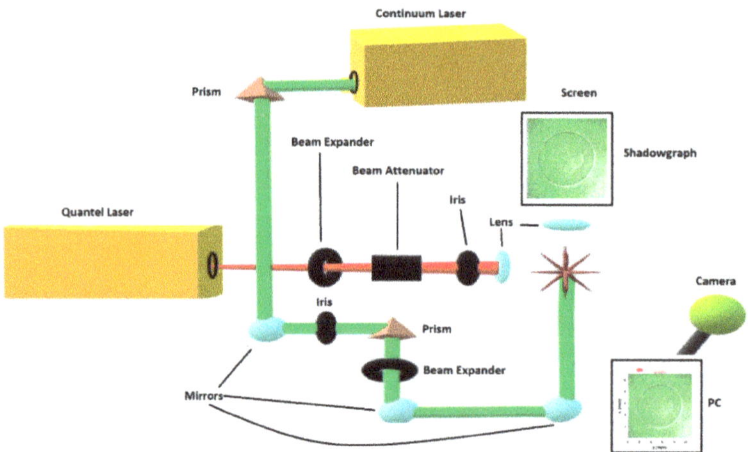

Figure 1. Schematic of the apparatus used for the shadowgraph experiments.

Synchronization of the lasers and camera (Silicon Video Color Camera model 9C10) were performed to ensure precise temporal measurements of shadowgraphs. A control box (TSI Incorporated model 610032, USA), a digital delay generator (Stanford Research System model DG535, USA), and two oscilloscopes were utilized in externally triggering the Surelite laser, Quantel laser, and camera to capture desired shadowgraphs. The externally triggered laser devices had 1 ns RMS amplitude trigger jitter.

Shadowgraphs were recorded using the silicon video camera equipped with a Nikon lens of 50 mm focal length and the silicon video camera was controlled by XCAP imaging software with a PIXCI image capture board. The silicon video camera is equipped with a sensitive area of 6.1 mm × 4.58 mm with a resolution of 3488 horizontal × 2616 vertical pixels or with square 1.75 μm × 1.75 μm pixels image resolution of 512 × 512 pixels was used to capture images by 4 × 4 grouping.

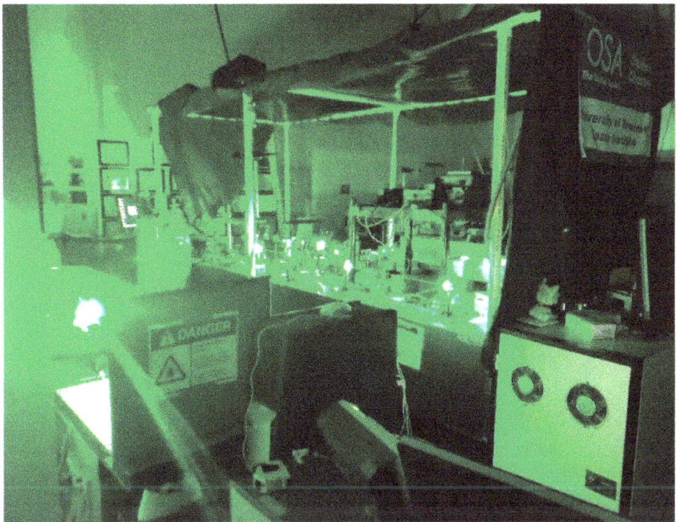

Figure 2. Photograph of the experimental arrangement for air breakdown.

2.2. Emission Spectroscopy

A set of typical components for nanosecond laser-induced breakdown spectroscopy (LIBS) or time-resolved laser-induced optical emission spectroscopy was used in the experimental arrangement [7]. A Q-switched neodymium-doped yttrium aluminum garnet, $Nd{:}Y_3A_{15}O_{12}$ (Nd:YAG) laser (Quantel model Q-smart 850, USA) is operated at a fundamental wavelength of 1064 nm and a 10 Hz repetition rate to deliver full-width-at-half-maximum (FWHM) 6 ns laser radiation with an energy of 850 mJ per pulse. Beam splitters and apertures were used to attenuate laser-pulse energy from 850 to 170 mJ per pulse. Energy per pulse is measured using the energy and optical power meter (Thorlabs model PM100USB, USA). A silicon photodiode detector (Thorlabs model DET10A/M, USA) is connected to a four- channel oscilloscope (Tektronix model TDS 3054, USA) to measure optical pulses from scattered laser radiation. Alignment of the laser beam is done with three IR reflective mirrors (Thorlabs model NB1-K13, USA) to ensure the laser beam is parallel to the spectrometer slit. A singlet lens (Thorlabs model LA-1509-C, USA) was used to produce optical-breakdown micro-plasma. A fused silica planoconvex lens (Thorlabs model LA4545, USA) was used for 1:1 imaging of the plasma onto the 100 μm slit of a Czerny–Turner type spectrometer (Jobin Yvon model HR 640, Fr) with a focal length of 0.64 nm. The Czerny–Turner type spectrometer has a spectral resolution of 0.1 nm and is equipped with a 1200 groove/mm grating to disperse the radiation from the plasma into numerous wavelengths. Figure 3 illustrates a schematic of the experimental setup.

Laser induced breakdown was created in a controlled gas chamber. The chamber consists of six 0.5 inch diameter and 1 cm thick uncoated uv-fs glass optical windows, two gas inlets, and one gas outlet. The gas inlets were used to mix Airgas ultra-high purity N_2 and research- grade CO_2 via partial pressure addition. Chamber evacuation was performed before and after experimentation via the gas outlet. A mechanical pump was connected to the gas outlet so the chamber could be evacuated as needed. To minimize contamination, the chamber was evacuated with a mechanical pump to bring the system pressure down to 1 Pa (10 mTorr). Chamber monitoring was performed using a differential pressure gauge. The gauge was operated as an absolute pressure gauge by evacuating the gauge casing volume via the vacuum pump. For these set of experiments, the chamber consisted of a 1:1 CO_2:N_2 molar gas mixture held at 101 kPa (760 Torr). Measurements were performed with and without an order-sorting filter (Oriel model 51250, USA) with a cut-on wavelength of 309 nm and transmittance range of 325 nm, to evaluate the C I 193.09 nm atomic carbon line interference [8]. Spatially resolved images were recorded with a 2-dimensional intensified charge coupled device (ICCD; Andor Technology model iStarDH334T-25U-04, USA) along the slit height. The ICCD is mounted to the spectrometer, connected to a computer by USB cable, and run by the Andor Solis 64-bit Acquisition and Analysis software. The ICCD has an array of 1024 × 1024 pixels, which coincide with horizontal-spectral and vertical-spatial variations along the slit height. The pixels are binned in four-pixel tracks along the slit direction, resulting in 256 spectra for each time delay. Recording of measurements with and without the order-sorting filter consist of 100 accumulations collected for 21 time-delays at 250 ns steps.

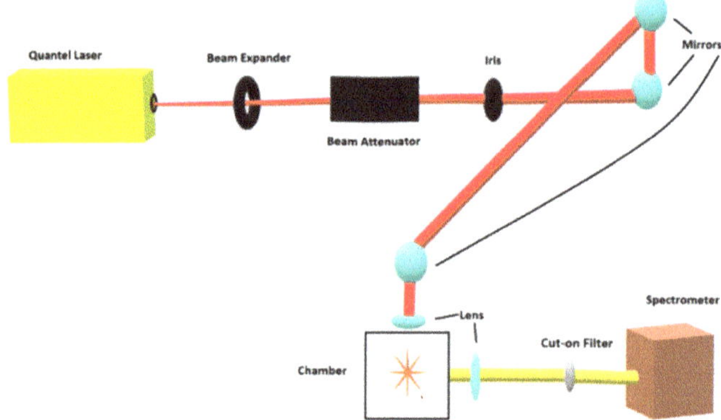

Figure 3. Schematic of the apparatus used for the laser-induced breakdown experiments.

Synchronization of plasma generating instrumentation and data collection equipment was accomplished by using a function generator (Wavetek model FG3C, USA), a digital delay generator (Stanford Research System model DG535, USA), a custom-built divide by five circuit, and the previously mentioned four-channel oscilloscope. In addition, 50 Hz triangular waves are produced by the wave generator and the divide by five circuit box yields a 10 Hz wave. The digital delay generator and production of 10 Hz wave are used to trigger the flashlamp of the laser and synchronize all other equipment.

2.3. *Shockwave Analysis Method*

Laser-induced breakdown performed on solid, liquid, or gaseous materials produces a small explosion. This explosion, together with the excitation, plasma formation, and ablation of material, is accompanied by the surrounding material being fiercely displaced and the production of a shockwave. The geometry and the evolvement over time of a laser-induced shockwave are dependent on the energy of the laser and the shape of plasma produced.

Blast waves due to nuclear explosions set the foundation for the study of shockwave production and propagation. The vast amounts of energy released in a fixed volume by nuclear explosions compared to normal explosions were examined by Bethe et al. [9] at Los Alamos, NM, USA, in 1941 and Taylor [10] in the United Kingdom in 1950, yielding the theory of a point strong explosion. Studies by John von Neumann [9] and L.I. Sedov [11,12], which assumed an adiabatic expansion and a sudden release of energy, E, in negligible volume and time, led to the development of the expansion law of the shock wave:

$$R(\tau) = \frac{1}{K}\left(\frac{E\,\tau^{2}}{\rho}\right)^{\frac{1}{5}}, \tag{1}$$

where $R(\tau)$ is the radius of the shockwave at time τ, K is a constant dependent on the adiabatic coefficient of the gas, and ρ is the density of the gas. For studies performed in standard ambient temperature and pressure (SATP) air, $K \approx 1$, which is consistent with shadowgraph studies performed by Gautam et al. [3]. Comparisons of computed blast-wave radii, using Equation (1) with K = 1, for SATP air and molar cyanide (CN) mixture are seen in Tables 1 and 2. There is minimal variation in shockwave expansion in SATP air versus CN mixture, which would indicate that measured shadowgraphs in air provide an acceptable guide for the CN gaseous mixture.

Table 1. Computed shockwave radii for SATP air and for molar CN mixture, 160 mJ.

τ (ns)	R (mm) for Air [ρ = 1.2 kg/m^3]	R (mm) for CN [ρ = 1.63 kg/m^3]
200	1.40	1.31
450	1.93	1.82
700	2.31	2.17
950	2.61	2.45
1200	2.86	2.69
1450	3.09	2.90

Table 2. Computed shockwave radii for SATP air and for molar CN mixture, 200 mJ.

τ (ns)	R (mm) for Air [ρ = 1.2 kg/m^3]	R(mm) for CN [ρ = 1.63 kg/m^3]
200	1.46	1.37
450	2.02	1.90
700	2.41	2.27
950	2.73	2.56
1200	2.99	2.81
1450	3.23	3.04

Another important characteristic of the shockwave is the expansion velocity. The shockwave expansion velocity indicates whether the approximation used in the shockwave expansion law is accurate. Shock wave expansion velocity, $v(\tau)$, assuming $K \approx 1$, is determined by:

$$v(\tau) = \frac{2}{5}\,\tau^{-3/5}\left(\frac{E}{\rho}\right)^{1/5}. \tag{2}$$

Studies by Harith et al. [13] discuss that the shockwave expansion law is a great approximation when the shockwave expansion velocity is around Mach numbers $Ma \approx 2$. However, this work also discusses applicability of the shockwave radius equation, Equation (1), for velocities with Ma >> 2, see Section 3.3.

Mach numbers, Ma, are calculated using:

$$Ma = \frac{v(\tau)}{c}, \tag{3}$$

where c is the speed of sound in SATP, 343 m/s. Comparisons of computed shock wave expansion velocities and Mach numbers for energies 160 and 200 mJ, using Equations (2) and (3), for SATP air and molar cyanide (CN) mixture are seen in Tables 3–6. Shockwave expansion velocities in air for early time delays (450 ns or less) move at hypersonic speeds (Mach numbers 5 or greater), while at later time delays (greater than 450 ns), the shock wave expansion velocities move at supersonic speeds (Mach numbers 1.3 to 5, inclusive). For the CN mixture, the shockwave expansion velocities move at supersonic speeds except for early time delays of 200 ns or less where they move at hypersonic speeds. Therefore, as time elapses, the shockwave expansion law approximation improves [13] as Mach numbers approach Ma $\approx$ 2 and slower.

Table 3. Computed shockwave velocity for SATP air and for molar CN mixture, 160 mJ.

τ (ns)	v (km/s) for Air [ρ = 1.2 kg/m^3]	v (km/s) for CN [ρ = 1.63 kg/m^3]
200	2.80	2.63
450	1.72	1.62
700	1.32	1.24
950	1.10	1.03
1200	0.95	0.90
1450	0.85	0.80

Table 4. Computed shockwave velocity for SATP air and for molar CN mixture, 200 mJ.

τ (ns)	v (km/s) for Air [ρ = 1.2 kg/m^3]	v (km/s) for CN [ρ = 1.63 kg/m^3]
200	2.92	2.75
450	1.80	1.69
700	1.38	1.30
950	1.15	1.08
1200	1.00	0.94
1450	0.89	0.84

Table 5. Computed shockwave Mach numbers for SATP air and for molar CN mixture, 160 mJ.

τ (ns)	Ma for Air [ρ = 1.2 kg/m^3]	Ma for CN [ρ = 1.63 kg/m^3]
200	8.15	7.67
450	5.01	4.71
700	3.84	3.61
950	3.20	3.01
1200	2.78	2.62
1450	2.48	2.34

Table 6. Computed shockwave Mach numbers for SATP air and for molar CN mixture, 200 mJ.

τ (ns)	Ma for Air [ρ = 1.2 kg/m^3]	Ma for CN [ρ = 1.63 kg/m^3]
200	8.52	8.02
450	5.24	4.93
700	4.02	3.78
950	3.35	3.15
1200	2.91	2.74
1450	2.60	2.44

2.4. Electron Density Determination Method

2.4.1. Atomic Carbon Line Interference

Laser-induced breakdown performed on the carbon dioxide and nitrogen gaseous mixture produces a variety of species, which includes C, C^+, C^-, CN^+, CN^-, CNN, CO, CO^+, CO_2, CO_2^+, C_2, C_2^+, C_2^-, CCN, CNC, C_2O, C_3, N, N^+, N^-, NCO, NO, NO^+, NO_2, N_2, N_2^+, N_2^-, NCN, N_2O, N_3, O, O^+, O^-, O_2, and O_2^+ [14]. This work focuses on the analysis of the CN violet-band $\Delta v = 0$ system that has vibrational bands (0, 0), (1, 1), (2, 2), (3, 3), and (4, 4), which are 388.34, 387.14, 386.19, 385.47, and 385.09 nm, respectively. The Czerny–Turner spectrometer equipped with 1200 groove/mm grating is adjusted to the region of interest of 370 to 393.5 nm to observe the CN violet band $\Delta v = 0$ system via the Andor Solis software [8].

In LIBS, first order lines ($m = 1$) are of interest when performing analysis of LIBS produced spectra, but spectral lines have different orders or modes due to the use of diffraction gratings which follow the equation:

$$d \sin(\theta) = m\lambda, \quad (4)$$

where d is the distance between the center of two adjacent slits, θ is the angle at which maxima occur, m is the propagation mode of interest, and λ wavelength of monochromatic light. For example, if a first order ($m = 1$) spectral line has a wavelength, λ, equivalent to 193 nm, then its wavelength, λ, measured in second order ($m = 2$) would be 386 nm when using Equation (4). Even though the spectrometer is set for the desired region of interest and the CN band heads are well defined, there can be interference from the other species' spectral lines in higher order that are produced by the laser-induced breakdown (LIBS). In the cyanide work, there is overlap of the CI 193.09 nm atomic carbon line in second order and the vibrational band (2, 2) of 386.19 nm. Measurements included recording of data without and with the previously mention 309 nm cut-on filter (see Section 2.2).

2.4.2. Line Broadening and Deconvolution

Spectral line broadening of observed plasma is caused by the effect of ions and electrons. Local conditions such as local thermodynamic equilibrium and extended conditions such as the plasma radiation's traversed path as viewed by an observer cause the spectral lines to broaden. Doppler or thermal broadening, natural broadening, and pressure broadening are different types of local effect broadening. Doppler broadening is characterized as a Gaussian profile and is due to the position of the detector or observer relative to the velocity of the atoms or ions within a gas or plasma. Natural broadening is characterized as a Lorentzian profile and occurs due to the uncertainty associated with the lifetime of the excited states and its energy. Lastly, pressure broadening is characterized as a Lorentzian profile and is caused by atoms or other ions neighboring the emitter atom or ion. One example of pressure broadening is Stark broadening, which is caused by the shifting and splitting of spectral lines due to the presence of an external electric field, known as the Stark effect.

Electron number density, n_e, can be determined from the full-width at half maximum (FWHM), $\Delta\lambda_{Stark}$, of the stark-broadened CI 193.09 nm atomic carbon line [15], measured in second order:

$$\Delta\lambda_{Stark}\ (nm) = 2w\ (nm) * n_e\left(10^{17}\ cm^{-3}\right), \quad (5)$$

where width parameter, w, was extrapolated [15,16], to be $w \approx 0.0029$ nm or by the Stark shift of the CI 193.09 nm atomic carbon line [15]:

$$\delta\lambda_{Stark}\ (nm) = d\ (nm) * n_e\left(10^{17}\ cm^{-3}\right), \quad (6)$$

where the shift parameter, d, was extrapolated to be [15,16] $d \approx 0.0029$ nm. In order to use Equations (5) and (6) to determine n_e, deconvolution of measured Stark FWHM and Stark shifts must be performed. This is due to the line broadening being largely influenced by Stark broadening, which is typically approximated using a Voigt profile [17]. The convolution of Gaussian and Lorentzian line profiles results in the Voigt profile, therefore using a rough approximation between Gaussian, Lorentzian, and Voigt profile widths:

$$f_V = \frac{f_L}{2} + \sqrt{\left(\frac{f_L}{2}\right)^2 + f_G^2}, \quad (7)$$

where f_V is the FWHM of the Voigt profile, f_L is the FWHM of the Lorentzian profile, and f_G is the FWHM of the Gaussian profile, can be used to apply deconvolution to the measured line profile. In this paper, f_V represents the measured spectral line, f_L represents the Stark FWHM or Stark shift, and f_G

represents the spectral resolution of the spectrometer. Therefore, rearranging Equation (7) to determine f_L yields:

$$f_L = f_V + \frac{f_G^2}{f_V}, \quad (8)$$

which can be used in conjunction with Equations (5) and (6) to determine n_e of the CI 193.09 nm atomic carbon line in second order.

2.4.3. Method for Computation of Electron Density

Spectra that were unfiltered had an overlap of the (2, 2) CN band head of 386.19 nm and the second order CI 193.09 nm atomic carbon line, where spectra filtered with the cut-on filter only had the (2, 2) CN band head of 386.19 nm. A peak-fitting Matlab® script [18] was applied to filtered and unfiltered spectra to evaluate CI 193.09 nm atomic carbon line in second-order Stark widths and shifts. A typical intermediate record of the five-peak fitting includes fitted background, a single profile for each of the five bands. Figure 4 illustrates recorded data and the overall 'peakfit.m version 9.0' result that is composed of background and sum of five Gaussians versus wavelength.

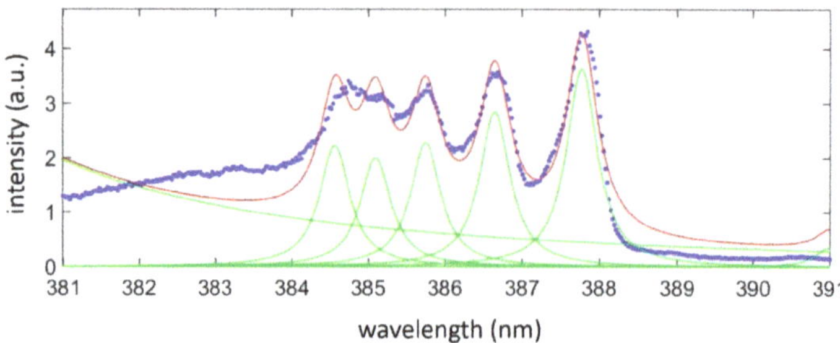

Figure 4. Typical result of the peakfit.m script applied to measured line-of-sight $\Delta v = 0$ CN spectra. Individual fitted peaks and the background variation (in green) are added up for the overall fit (in red) to the experimental data (dotted, in blue).

Equation (8) was used on the extracted Stark full-width at half maximum (FWHM) and Stark shifts to apply deconvolution. The difference between deconvoluted Stark FWHM from unfiltered spectra and deconvoluted Stark FWHM from filtered spectra was performed to determine the FWHM contribution of the CI 193.09 nm atomic carbon line in second order only. The FWHM contribution of the CI 193.09 nm atomic carbon line in second order only was used in conjunction with Equation (5) to determine n_e, where deconvoluted Stark shifts can be used in conjunction with Equation (6) to determine n_e as well. This communication reports results of n_e from only Stark widths.

2.5. Molecular Spectra Analysis Method

Plasma temperature from molecular spectra is dependent on vibrational and rotational elements of the molecular spectrum. Due to this dependence, temperature from molecular spectra can be used to evaluate the condition of the plasma. Temperature determination can be performed by fitting measured spectra to calculated theoretical diatomic spectrum. Diatomic line strength is used to calculate the needed theoretical diatomic spectrum for appropriate fitting. Diatomic line strength calculations are rather cumbersome to perform due to spectral line position requirements and the need for very accurate molecular rotational constants. Parigger and Hornkohl [19] describe the theoretical background and development of diatomic line strength tables necessary for temperature evaluation. Therefore, cyanide (CN) line strength tables reported by Parigger et al. [20] are used for the calculation of theoretical CN spectrum.

The Nelder–Mead temperature (NMT) program [17] was used to fit measured CN spectral data. The NMT program utilizes the Nelder–Mead method [21], which is a numerical method used to find the minimum or maximum of an objective function in a multidimensional space. The NMT program requires initial fit parameter assumptions, which consist of the temperature of the molecular spectra, the line width of the molecular spectral line, and a linear baseline offset. Specifically, the Nelder–Mead algorithm creates a simplex established on the initial given fit parameters. Each vertex of the simplex represents a fit parameter and the size of the simplex is reduced by changing the vertex arrangement until the tolerance is achieved. During minimization, the first local minimum identified represents the minimum and the best fit parameters are established on final location of the simplex's vertices.

2.6. Abel Inversion Method

The Abel transform, which analyzes spherically symmetric functions, can be applied to evaluate line-of-sight measurements from close to spherically symmetric plasmas. Figure 5 illustrates the line-of-sight experimental geometry. Specifically, the Abel inversion [22–26] technique allows one to extract the radial distributions of electron densities of a close to spherically symmetric plasma directly from the recorded line-of-sight data. Letting ε(r, λ) represent the radial emission coefficient, the Abel transform of ε(r, λ) is shown to be [5,8,14,27–30]:

$$I(z, \lambda) = 2\int_{z}^{P \gg R} \varepsilon(r, \lambda)\frac{r}{\sqrt{r^2 - z^2}}\,dr, \tag{9}$$

where z is the perpendicular distance from the origin of the line-of-sight, r is the radial distance from the center of the observed plasma at which electron density will be evaluated, R is the radius of the spherical object and P is the upper integration limit which is established as being much greater than R, P >> R. The recorded emission-intensity map contains spatial information along the slit dimension, and spectral information along the wavelength dimension. Abel inversion is performed on measured line-of-sight data of the emitted spectral intensity in order to determine the radial variation.

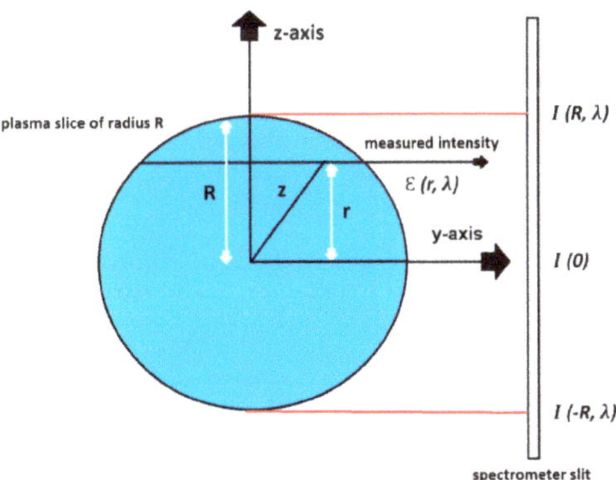

Figure 5. Line-of-sight geometry and Abel inversion method.

The emission intensity has subtle asymmetry along the direction perpendicular to the laser beam which may be a result of laser-plasma interaction that normally takes place in nanosecond laser-induced plasmas or may be a result of variations of the laser pulse profile. Figure 5 displays the geometry of the line-of-sight measurements utilized in this work, where line-of-sight measurements are along the *y*-axis, the direction of the laser beam is along the *z*-axis, and the *x*-axis is perpendicular to the y and z axes. Abel inversion is used to obtain radial dimension measurements of the plasma, which allows for the extraction of plasma radial information. Using the top half and bottom half of line-of-sight profiles, asymmetric data points are averaged utilizing the same symmetrization method for atomic hydrogen spectra [27–29] and application of Abel inversion is used.

Shadowgraph measurements of plasma kernels in hydrogen and hydrogen–nitrogen gaseous mixtures show a close to cylindrical symmetry or prolate spheroidal symmetry as compared to the close to spherically symmetric plasma kernels in standard ambient temperature and pressure in laboratory air for time delays of 100 to 10,000 ns [3,28–30]. The deviation from close to spherically symmetric plasma kernels may be due to the laser energy employed for laser-induced breakdown, therefore as discussed in this work, shadowgraph measurements in standard ambient temperature and pressure are performed at laser energies observed in the carbon dioxide and nitrogen gaseous mixture, which are seen to be close to spherically symmetric [5]. Due to the spherically symmetric plasma requirement of the Abel inversion technique, collected line-of-sight data are adjusted to allow subtle deviations from circular symmetry and modeled to be spherically symmetric [27–29].

Abel inversion of each wavelength determines the spatial distribution and subtle asymmetries present in the captured data are kept by applying an asymmetric factor, which then establish the asymmetric radial distributions [27–29]. Pre-treatment of the captured data is not required when using the derivative free Abel inverse transform method [22]. Inversion of Equation (9) can be accomplished using an analytical method with derivatives, known as the Abel inverse transform. Differentiation of spectra is rather challenging; therefore, coefficients are determined by using a complete set of orthogonal polynomials with a minimization method. The use of Chebyshev polynomials in conjunction with the available Matlab® script [22,26] for Abel inversion of Equation (9), allows the direct inversion of measured data. For this work, inversion was accomplished by choosing 15 polynomials [22,23], which maintained fidelity of the spectra and was comparable to the use of a digital filter resulting in computed radial spectra. Smaller spectral resolution would occur with the selection of a smaller number of polynomials. Line-of-sight data along the spectrometer slit were captured and inverted for each wavelength to get the radial intensity distribution. Calibration and correction for system sensitivity using standard lamps is required for recorded data to undergo Abel inversion and curation of the spectra.

3. Results

3.1. Shadowgraphs

Shadowgraphs of plasma in standard ambient air and temperature (SATP) produced by infrared (IR) 1064 nm radiation with excitation energy of 170 mJ and 6 ns pulse-width are shown below. A single 5 ns pulse-width 532 nm laser beam was used to capture the shadowgraphs. Shadowgraphs were taken in the range of 0.2 to 4.2 µs time delays. Figure 6 displays typical results.

Further investigations of laser-induced laboratory air breakdown utilize pulse energies of 850 mJ per 6 ns, 1064 nm pulses. Figures 7 and 8 illustrate recorded images in the range of 200 to 4000 ns.

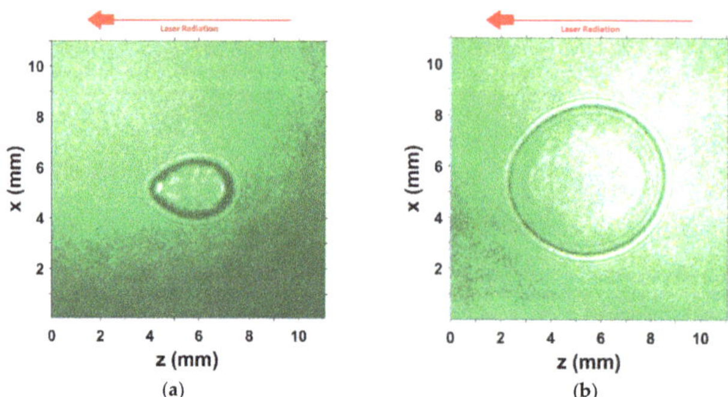

Figure 6. *Cont.*

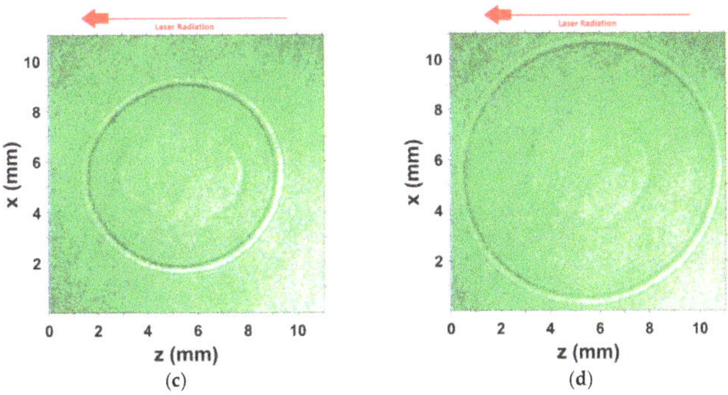

Figure 6. Single-shot shadowgraphs of expanding laser-induced plasma initiated with 170 mJ, 6 ns, 1064 nm pulses, and imaged with 5 ns, 532 nm backlight, time-delayed by (**a**) 200 ns; (**b**) 1200; (**c**) 2200 ns; (**d**) 4200 ns.

Figures 7 and 8 display multiple breakdown events along the optical axis. Stagnation layers appear to be formed between individual breakdown spots, developing into vertical structures in the forward direction. Stagnation layers have been explored at the interface region of two colliding laser-induced plasmas [31]. The predicted initial plasma expansion speeds in laser-induced optical breakdown are of the order of 100 km/s [32], followed by a gas expansion that is analogous to that of a strong explosion [32]. Figures 7 and 8 also exhibit associated early expansion dynamics that occur at speeds well in excess of hypersonic speeds. The blue lines indicate the forward propagating shockwave boundaries that originate from multiple breakdown spots appearing as 'beads'.

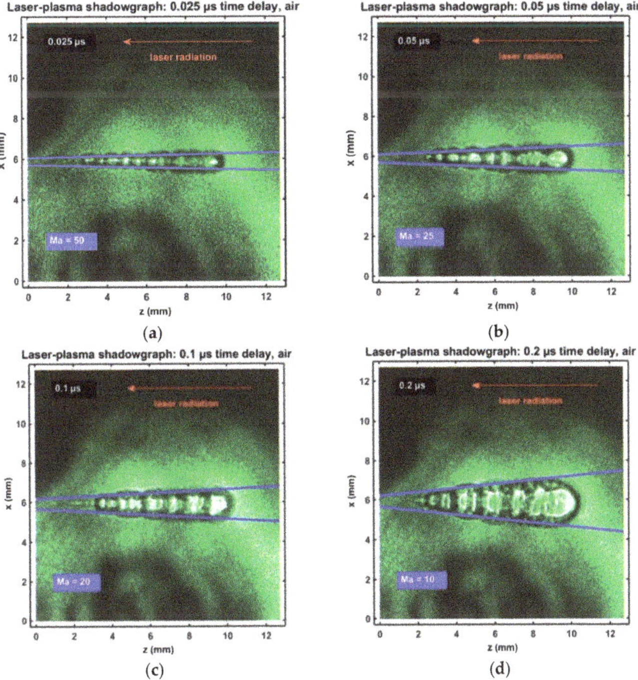

Figure 7. *Cont.*

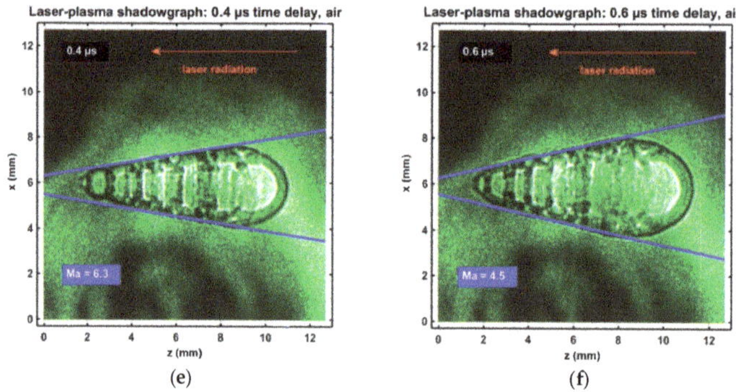

(e) (f)

Figure 7. Shadowgraphs subsequent to laser-plasma generation with 850 mJ, 6 ns, 1064 nm pulses. Time delays: (**a**) 25 ns; (**b**) 50 ns; (**c**) 100 ns; (**d**) 200 ns; (**e**) 400 ns; (**f**) 600 ns.

As indicated in Figure 6, the IR 1064 nm, 170 mJ, 6 ns laser beam is along the z axis and moving from the right to left. The expanding shockwave and plasma kernel are clearly visible. At 0.2 µs delay, the plasma kernel appears cylindrical and the expanding shockwave has a prolate spheroidal shape. As time delays approach 1 µs, the plasma kernel and expanding shockwave become nearly spherical. As time elapses further, the plasma kernel and expanding shockwave continue to become close to spherical. The vertical extend is about a factor of 1.4 smaller for 170 mJ pulses than that for 850 mJ pulses, according to the Taylor–Sedov energy dependency, Equation (1), for spherical expansion.

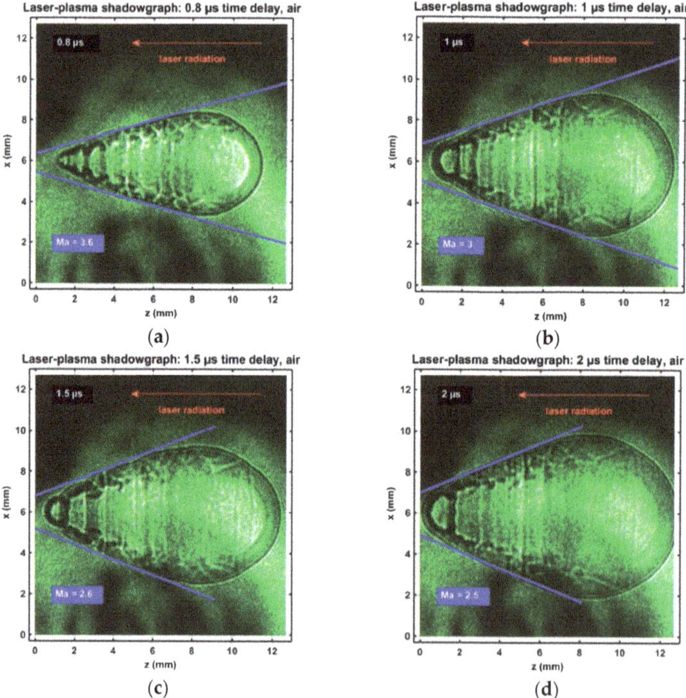

(a) (b) (c) (d)

Figure 8. *Cont.*

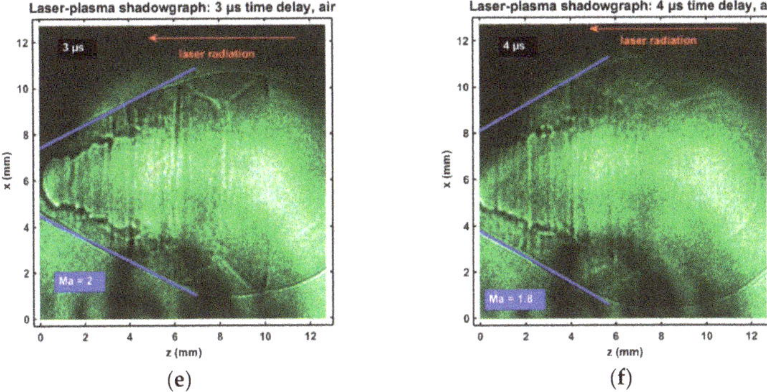

(e) (f)

Figure 8. Shadowgraphs captured after laser-plasma generation with 850 mJ, 6 ns, 1064 nm pulses. Time delays: (**a**) 800 ns; (**b**) 1000 ns; (**c**) 1500 ns; (**d**) 2000 ns; (**e**) 3000 ns; (**f**) 4000 ns.

3.2. Emission Spectra

The cyanide (CN) spectra captured by the spectrometer and 2-dimensional intensified charge coupled device for a fixed volume of the 1:1 molar CO_2:N_2 gas mixture held at atmospheric pressure are shown in Figure 9. As seen in Figure 9, the CN violet system $B^2\Sigma^+ - X^2\Sigma^+$ vibrational bands of (0, 0), (1, 1), (2, 2), (3, 3), and (4, 4) are clearly visible and discernible. The overlap of the CI 193.09 nm atomic carbon line in second order and the (2, 2) CN band head is also seen in Figure 9. At time delays greater than 2.5 μs, the CI 193.09 nm atomic carbon line in second order appears to dissipate and does not overlap the (2, 2) CN band head.

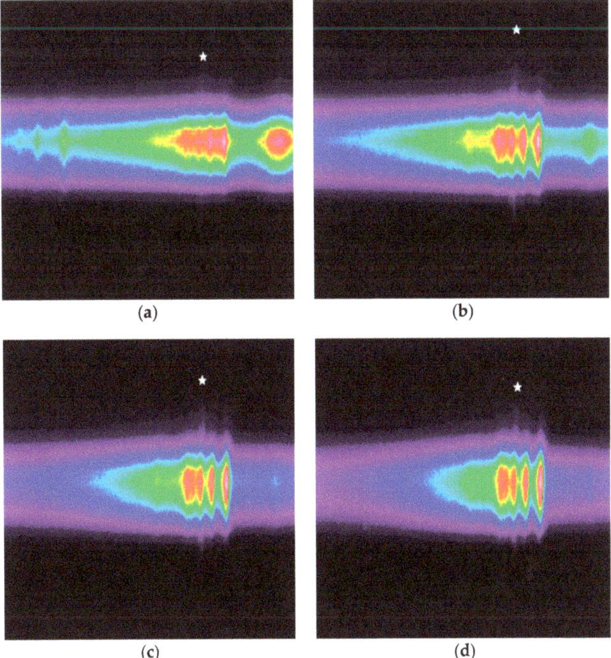

(a) (b) (c) (d)

Figure 9. Optical breakdown CN spectra in a 1:1 molar CO_2:N_2 gas mixture held at atmospheric pressure for time delays of (**a**) 200 ns, (**b**) 450 ns, (**c**) 700 ns, and (**d**) 950 ns. Spectrometer-detector gate width: 125 ns. *, second-order atomic carbon line [5].

In separate experimental runs, CN spectra were captured for a fixed volume of the 1:1 molar CO_2:N_2 gas mixture held at atmospheric pressure with the use of a 309 nm cut-on wavelength filter. The 309 nm cut-on wavelength filter allows for the suppression of the CI 193.09 nm atomic carbon line in second order. Although it is advantageous to apply the 309 nm cut-on wavelength filter for the suppression of the CI 193.09 nm atomic carbon line in second order, the 309 nm cut-on wavelength filter causes a reduction in spectral intensity captured by the spectrometer and ICCD by ≈ 13%. Filtered and unfiltered spectra also show the CN plasma moving towards the laser as time elapses.

3.3. Shockwave amd Plasma Expansion

The expanding shockwave radii results are shown in Table 7, decreasing shockwave velocities and Mach number results are shown in Table 8, and increasing plasma kernel radii are shown in Table 9. The expanding shockwave radius for 0.2 μs delay is not exactly consistent with the previously discussed shockwave expansion law, Equation (1), and this can be due to the velocity of the shockwave being greater than the Mach 2 maximum velocity requirement of the shockwave expansion law. For time delays approaching 1 μs and later, the expanding shockwave radii are consistent with the shockwave expansion law, with their shockwave expansion velocities, v(τ), being closer to Mach 2 and slower.

Table 7. Computed shockwave radii versus measured shockwave radii for SATP air, 170 mJ.

τ (ns)	Computed R (mm)	Measured R (mm)
200	1.41	1.00 ± 0.30
1000	2.69	2.67 ± 0.80
1200	2.90	2.83 ± 0.85
2200	3.69	3.57 ± 1.07
4200	4.78	4.95 ± 1.49

Table 8. Inferred shockwave velocities and Mach numbers for SATP air, 170 mJ.

τ (ns)	Velocity, v (km/s)	Mach number, Ma
200	4.03 ± 1.21	11.76 ± 0.30
1000	1.31 ± 0.39	3.82 ± 1.15
1200	1.08 ± 0.32	3.15 ± 0.95
2200	0.58 ± 0.17	1.67 ± 0.50
4200	0.30 ± 0.09	0.87 ± 0.26

Table 9. Measured plasma kernel radii for SATP air, 170 mJ.

τ (ns)	Measured r (mm)
200	0.45 ± 0.13
1000	2.25 ± 0.67
1200	2.40 ± 0.72
2200	3.00 ± 0.90
4200	4.04 ± 1.21

As previously discussed, shockwave expansions in SATP air appear similar in the CN mixtures used in this work. Therefore, the visualization of these shockwave expansions in SATP air provides a good model to the shockwave expansion behavior and plasma kernel geometry in the CN mixtures.

The images recorded for 850 mJ laser-induced plasma generation are also analyzed using the Taylor–Sedov theory of blast wave propagation from a point explosion yields the time dependent radius, R(t), of the shock front [27–29]:

$$R(\tau) = \xi\,(E/\rho)^{\frac{1}{n+2}}(\tau)^{\frac{2}{n+2}} \sim (\tau)^{\frac{2}{n+2}}. \tag{10}$$

Here, ξ (ξ = 1/K) is a constant in the range of 1.0 to 1.1 that depends on the specific heat capacity, E is the energy that is released during the explosion or the absorbed energy per laser pulse, ρ is the gas density, τ is the time delay, and n is the shape dependent parameter. The values of n = 1, 2, 3 correspond to planar, cylindrical, and spherical shock waves, respectively.

One can use Equation (10) for computation of the blast wave or shock front expansion generated from laser-induced optical breakdown. However, of primary interest is the dependence of the radius, R(τ), on time delay, τ. Figure 10 displays the maximum of the shock wave radius versus time delay measured perpendicular to the direction of the laser beam propagation. In view of Figures 8 and 9, the maximum is determined from the images near z = 8 mm.

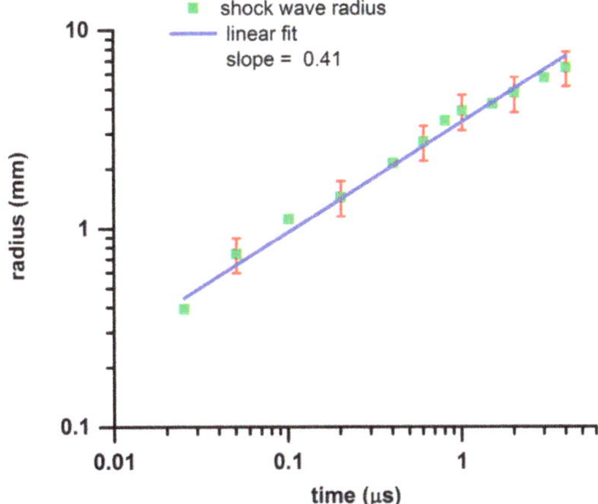

Figure 10. Log-log plot of shock wave expansion measured perpendicular to the laser-propagation direction when using 850 mJ, 6 ns, 1064 nm pulses for optical breakdown in laboratory air [4].

The linear fit (Figure 10) reveals 0.41 for the slope, or $n = 2.9{\sim}3$. In other words, spherical expansion is inferred. The figure also shows 20% error bars. These error bars are estimated from the variations in the pulse energy for generation of optical breakdown, the trigger-jitter synchronization of the two laser beams (one for plasma generation, the other from a separate device for shadowgraphs), and the readout errors from the displayed images in Figures 2–4. One can also extract from the graph the approximate 1 mm per μs expansion velocity for time delays of ~1 μs, or Ma = 3. From Equation (10), using $\xi = 1.0$ to 1.1, $E = 800$ mJ, $\rho = 1.225$ kg/m^3, and $n = 3$ yields for the radius R(τ = 1 μs) = 3.7 to 4.1 mm, consistent with the measured value of 3.9 mm [4].

For ultra-high-pure hydrogen, and for time delays of the order of almost 0 to a few dozen ns, recorded spectral images are utilized for exploration of the well-above hypersonic expansion. Figure 11 illustrates two images captured from optical breakdown in near atmospheric hydrogen gas, i.e., at a cell pressure of (1.08 ± 0.033) × 10^5 Pa (810 ± 25 Torr).

Figure 12 summarizes the expansion speed for early time delays. However, the speed of sound in hydrogen is approximately 4× higher than in air, and the recorded air shadowgraphs can serve as a guide for shockwave appearances. For example, hydrogen expansion at a delay of 400 ns approximately corresponds to air shadowgraphs recorded at a delay of 1600 ns. Most importantly, if the irradiance is not significantly higher than that for optical breakdown thresholds, a spherically symmetric appearance of the shockwave for delays 10× larger than indicated in Figure 12 would be expected analogous to the 170 mJ shadowgraphs recorded in air. Indeed, captured shadowgraphs of optical breakdown in hydrogen [30] reveal a prolate spheroidal shockwave shape for a time delay of 400 ns (see Figure 1 in Ref. [30]).

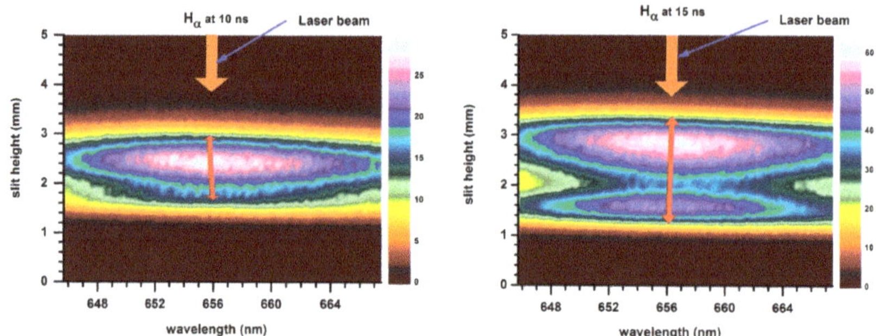

Figure 11. Hydrogen alpha plasma spectra images at 10 ns (left) and 15 ns (right) time delays. The red arrow indicates the measured plasma width [4].

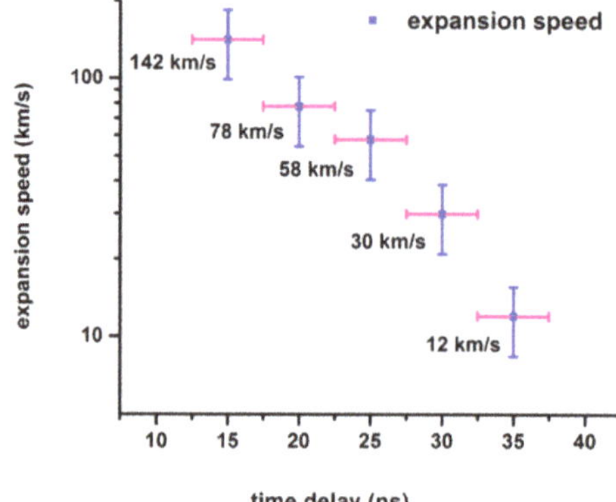

Figure 12. Plasma expansion speeds. The indicated time-delay error bars are due to the gate width of 5 ns [4].

The determined expansion speeds for hydrogen shockwave expansion speeds are well-above hypersonic speed (hypersonic: Ma $\geq$ 5) or above re-entry speeds (re-entry: Ma $\leq$ 25) at time delays of 10 to 40 ns.

3.4. Electron Density

The inferred Stark widths of the CI 193.09 nm carbon line in second order for the 1:1 molar CO_2:N_2 gaseous mixture held at atmospheric pressure were determined using the previously discussed peak-fitting Matlab® script [18], deconvolution of the filtered and unfiltered measured peaks, and taking the difference between the filtered and unfiltered deconvoluted peaks.

The inferred Stark widths are plotted versus the slit height of the spectrometer, which can be seen in Figure 13. Larger Stark widths are seen towards the edges of the plasma, while smaller Stark widths are seen in the center of the plasma.

For a time delays of 450 and 950 ns, Figure 13a,c, the Stark widths are between 0.4 and 0.5 nm and located towards the edges of the plasma. The Stark widths are used to calculate electron number density, n_e. The calculated n_e is plotted versus the slit height of the spectrometer, which can be seen in Figure 13b,d. Peak electron densities are of the order of $n_e \approx 10^{19}$ cm^{-3}, and values between the two peaks are of the order of $n_e \approx 10^{17}$ cm^{-3}. Since n_e is directly proportional to the Stark width of the 193.09 nm carbon line in second order as previously shown in Equation (5), n_e plots mimic the same behavior as the previously mentioned Stark width plots, where higher n_e is seen towards the edges of the plasma and lower n_e is towards the center of the plasma. The higher electron densities toward the edges of the plasma appear to follow the previously discussed shockwave expansion law, Equation (1), within the indicated error bars as seen in Table 10.

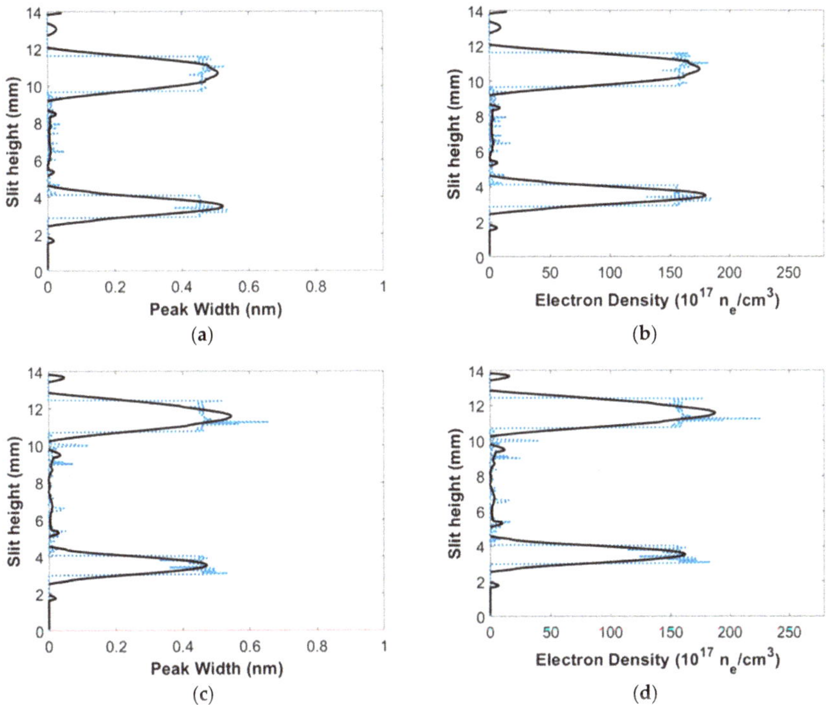

Figure 13. Inferred widths and calculated electron densities of C I 193.09 nm atomic carbon line in 2nd order vs. slit height for 1:1 molar CO_2:N_2 gas mixture held at atmospheric pressure. Time delays: (**a**,**b**) 450 ns, and (**c**,**d**) 950 ns.

Table 10. Computed shockwave radii versus plasma radius for 1:1 molar CO_2:N_2 gaseous mixture held at atmospheric pressure, 170 mJ.

τ (ns)	Computed R (mm)	Measured R_{Plasma} (mm)
450	1.84	2.90 ± 0.87
700	2.20	3.00 ± 0.90
950	2.48	3.30 ± 0.99
1200	2.72	3.35 ± 1.01
1450	2.94	4.00 ± 1.20
1700	3.13	4.30 ± 1.29
1950	3.31	4.40 ± 1.32
2200	3.47	4.30 ± 1.29

The Stark shifts of the CI 193.09 nm carbon line in second order for the 1:1 molar CO_2:N_2 gaseous mixture held at atmospheric pressure were also determined using the peak-fitting Matlab® script [18]. Larger Stark shifts are seen towards the edges of the plasma, while smaller Stark widths are seen in the center of the plasma. However, similar results are found when using the Stark shifts, yet with the shock wave fronts more precisely demarcated when using the Stark widths.

3.5. Cyanide Temperature

Inferred temperatures of filtered line-of-sight CN spectra in the 1:1 molar CO_2:N_2 gaseous mixture held at atmospheric pressure are plotted versus slit height of the spectrometer as shown in Figure 14. The outgoing shockwave can be seen from time delays of 450 to 950 ns. Figure 14 indicates that temperature variations occur in the central region, while increased temperatures are shown at the edges of the plasma. Higher temperatures are seen on the edge of the plasma towards the top of slit or towards the laser side. At a time-delay of 450 ns, Figure 14a, the temperatures in the central region of the plasma are between 9500 and 10,000 K, while the temperatures at the edges of the plasma are more than 10,000 K. At time delays of 950 ns, Figure 14b, the temperatures in the central region of the plasma cool to a range of 9000 to 9500 K, while temperatures at the edges of the plasma are between 9500 and 10,000 K. As time elapses further the plasma central region, temperatures cool even further to a range of 8500 to 9000 K for time delays of 1.2 to 1.7 µs, while the edges of the plasma maintain a temperature range of 9500 to 10,000 K. From time delays of 1.95 to 2.2 µs, the central region of the plasma sustains temperatures of 8500 to 9000 K and temperatures near the edge of plasma towards the bottom of the slit are around 9000 K, while temperatures near the edge of the plasma towards the top of the slit increase to greater than 11,000 K.

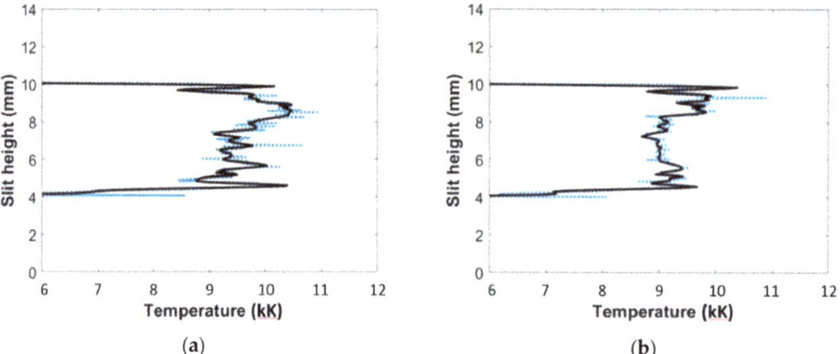

Figure 14. Temperature vs. slit height for filtered line-of-sight CN spectra for fixed volume of 1:1 molar CO_2:N_2 gaseous mixture held at atmospheric pressure. Time delays: (**a**) 450 ns; (**b**) 950 ns.

3.6. Abel Inverted Spectra

Abel inversion of the filtered 1:1 molar CO_2:N_2 gaseous mixture held at atmospheric pressure was performed by inverting measured line-of-sight data, I(z, λ), for each wavelength, λ, to obtain the radial emissivity distribution, ε(r, λ). Figures 15 and 16 illustrate rsults. Previously measured shadowgraphs show the plasma generated in the 1:1 molar CO_2:N_2 gaseous mixture held at atmospheric pressure has a close to spherical shape, which would justify the use of Abel inversion.

The irradiance threshold for optical breakdown for the experimental arrangement (see Section 2.1) is ≈ 20 mJ [5], or in terms of irradiance $\approx 3 \times 10^{11}$ W/cm^2. When using pulse energies of up to about 170 mJ, or up to ≈ 10 × breakdown threshold, close to symmetric shock wave expansion occurs. For pulse energies of 800 mJ, or 40 × breakdown threshold, the "symmetric" expansion is seen in the region where most of the energy is absorbed (see Figures 7 and 8).

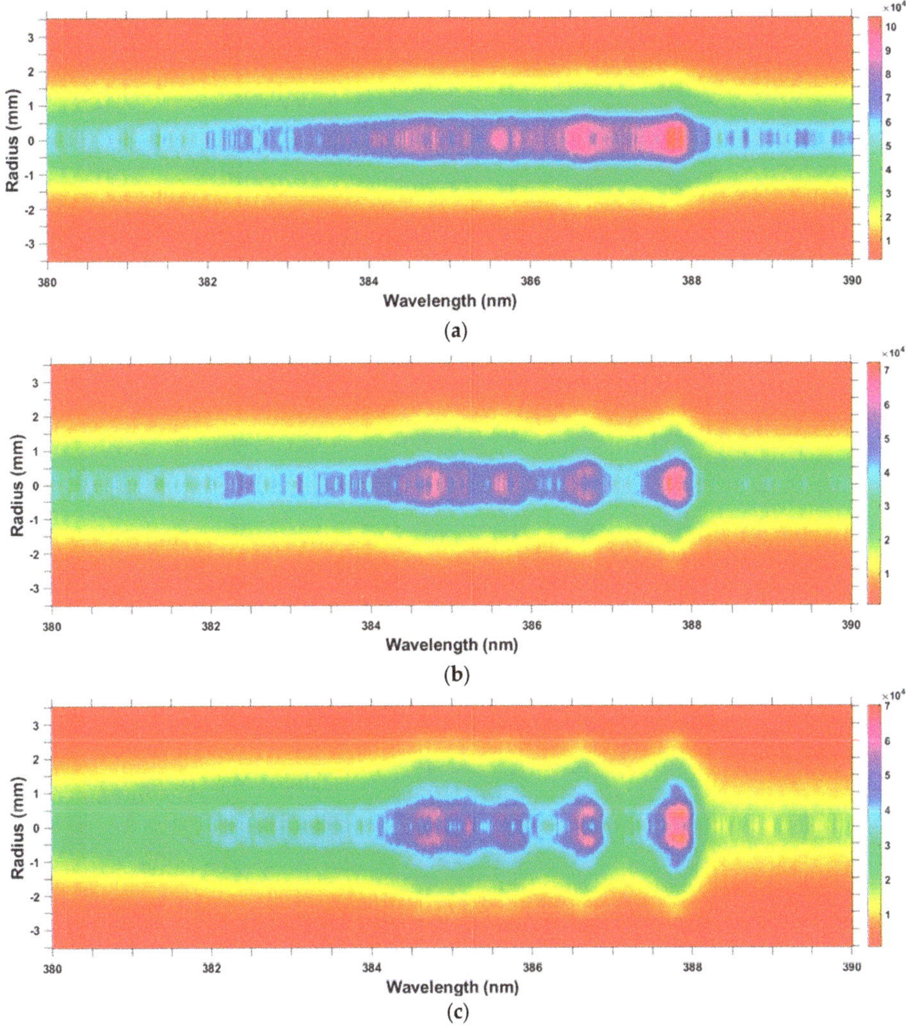

Figure 15. Abel inverted CN spectra 1:1 molar CO_2:N_2 gaseous mixture held at atmospheric pressure. Time delay: (**a**) 200 ns; (**b**) 450 ns; (**c**) 950 ns.

At a time-delay of 200 ns, Figure 15a, the CN distribution appears evenly distributed across the plasma. From time delays of 450 to 2200 ns, Figure 15b,c and Figure 16a–c, the CN signals begin to become stronger towards the edges of the plasma and weaker in the center of the plasma, which is consistent with the higher temperatures seen at the edges of the plasma as discussed previously. These results were projected to be similar to the shockwave results, but inside the plasma kernel and shockwave, the variations of the CN distribution were expected as well.

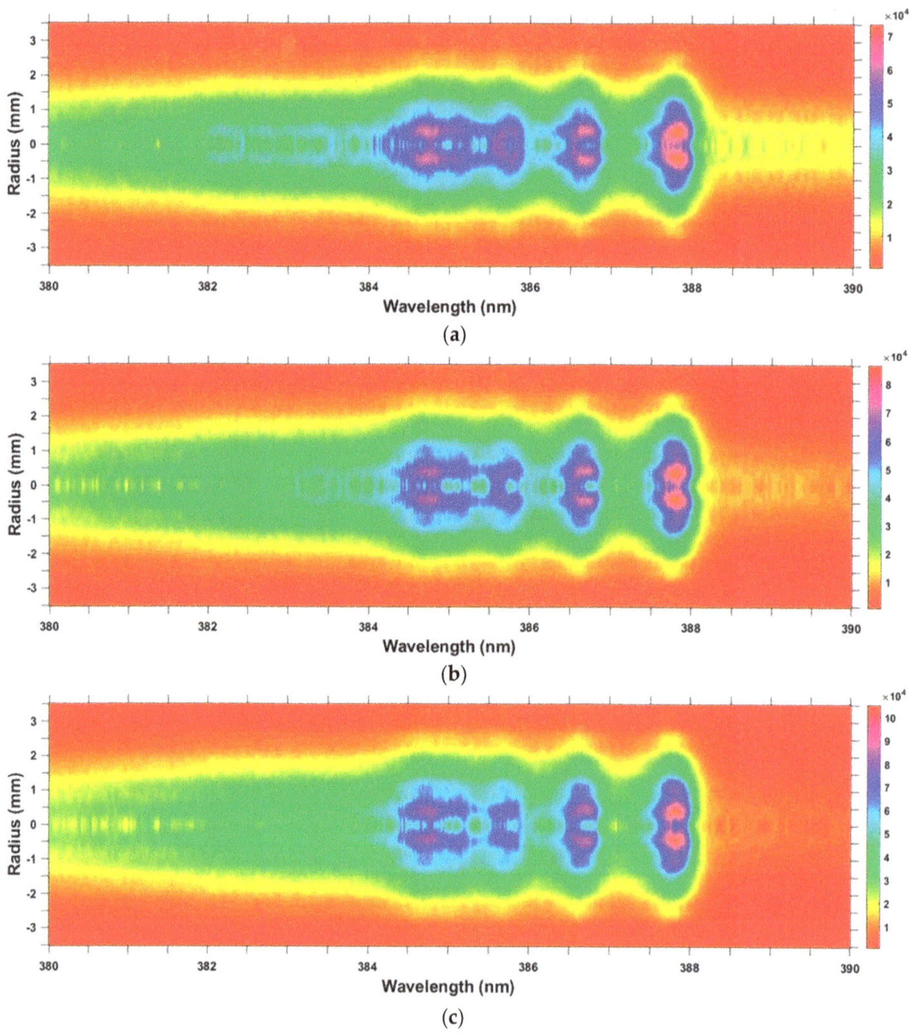

Figure 16. Abel inverted CN spectra 1:1 molar CO_2:N_2 gaseous mixture held at atmospheric pressure. Time delay: (**a**) 1200 ns; (**b**) 1700 ns; (**c**) 2200 ns.

4. Discussion and Conclusions

The laser-induced optical breakdown studies of air and selected gases reveal that usually multiple breakdown spots occur along the optical axis when focusing radiation to above threshold irradiance. Once optical breakdown is achieved, the absorbed radiation energy drives the shockwave towards the incoming radiation. The forward cone, or the initial asymmetry noticeable in the shadowgraphs, is a measure of how high above-threshold irradiance is employed.

The species concentration near the shockwave is higher than in the center, especially well developed for time-delays of the order of 1 μs. Both atomic species and diatomic molecular species such as CN indicate consistent results. Moreover, comparison of line-of-sight and of Abel-inverted data show agreement, especially at the spatial location of a breakdown 'bead' closest to the incoming beam one notices the development of the laser-induced plasma that is close to spherical for irradiance levels about one order of magnitude higher than threshold. For higher irradiances, the shockwave appears to expand into spherical shape towards the laser-side, in agreement with computed spherically expanding shockwaves.

The presented investigations of cyanide formation, especially near the expanding shockwave, are instrumental for potential medical and industrial cyanide diagnosis applications. But clearly, CN detection with LIBS shows the following conclusions:

- Shockwave expansion affects the formation of CN molecules as the plasma expands;
- Stark widths and shifts can be used to determine electron density, but higher spectral resolutions would be desirable for determination of accurate values of electron densities;
- For time delays around 1 μs, higher CN and electron concentrations occur near the shockwave than those in the central region of the plasma. The CN becomes concentrated towards the edges of the plasma, therefore slit size, energy per pulse, and measurement acquisition time would need to be considered when capturing data especially for handheld design;
- The use of a 309 nm cut-on filter is an effective way to filter out unwanted atomic carbon line contributions but causes a ~10% reduction in the signal captured, which can cause issues with possible quantification for medical and forensic applications;
- As plasma expands and cools, radiation from excited CN molecules seems evenly distributed and indicates a close to homogenous temperature;
- Abel inversion is only justified for radially symmetric light sources, but shadowgraph studies support symmetrization to elucidate spatial dependence.

Author Contributions: Data curation, C.G.P., C.M.H. and G.G.; Formal analysis, C.G.P., C.M.H. and G.G.; Investigation, C.G.P., C.M.H. and G.G.; Visualization, C.G.P., C.M.H. and G.G.; Writing – original draft, C.G.P., C.M.H. and G.G.; Writing – review & editing, C.G.P., C.M.H. and G.G. All authors have read and agreed to the published version of the manuscript.

Funding: This research received no external funding.

Acknowledgments: The authors wish to acknowledge the support in part by the Center for Laser Applications at the University of Tennessee Space Institute.

Conflicts of Interest: The authors declare no conflict of interest.

References

1. Miziolek, A.; Palleschi, V.; Schechter, I. (Eds.) *Laser Induced Breakdown Spectroscopy (LIBS): Fundamentals and Applications*; Cambridge University Press: Cambridge, UK, 2006.
2. Singh, J.P.; Thakur, S.N. (Eds.) *Laser-Induced Breakdown Spectroscopy*, 2nd ed.; Elsevier: Amsterdam, NL, USA, 2020.
3. Gautam, G.; Helstern, C.M.; Drake, K.A.; Parigger, C.G. Imaging of laser-induced plasma expansion dynamics in ambient air. *Int. Rev. At. Mol. Phys.* **2016**, *7*, 45–51.
4. Gautam, G.; Parigger, C.G. Plasma Expansion Dynamics in Hydrogen Gas. *Atoms* **2018**, *6*, 46. [CrossRef]
5. Parigger, C.G.; Helstern, C.M.; Jordan, B.S.; Surmick, D.M.; Splinter, R. Laser-plasma spatio-temporal cyanide spectroscopy and applications. *Molecules* **2020**, *25*, 615. [CrossRef] [PubMed]
6. Helstern, C.M. Laser-Induced Breakdown Spectroscopy and Plasmas Containing Cyanide. Ph.D. Thesis, University of Tennessee Knoxville, Knoxville, TN, USA, 2020.
7. Cremers, D.A.; Radziemski, L.J. *Handbook of Laser-Induced Breakdown Spectroscopy*; John Wiley and Sons: Hoboken, NJ, USA, 2006.
8. Parigger, C.G.; Helstern, C.M.; Gautam, G. Molecular emission spectroscopy of cyanide in laser-induced plasma. *Int. Rev. At. Mol. Phys.* **2017**, *8*, 25–35.
9. Bethe, H.A.; Fuchs, K.; Hirschfelder, J.O.; Magee, J.L.; Peierls, R.E.; Neumann, J. *Blastwave*; University of California: Los Alamos, NM, USA, 1947.
10. Taylor, G.I. The formation of a blast wave by a very intense explosion i. Theoretical discussion. *Proc. Math. Phys. Eng. Sci.* **1950**, *201*, 159–174.
11. Sedov, L.I. *Similarity and Dimensional Methods in Mechanics*; Academic Press: Cambridge, MA, USA, 1959.

12. Sedov, L.I. Propagation of strong blast waves. *Prikl. Mat. Mech.* **1946**, *10*, 241–250.
13. Harith, M.A.; Palleschi, V.; Salvetti, A.; Singh, D.P.; Tropiano, G.; Vaselli, M. Hydrodynamic evolution of laser driven diverging shock waves. *Laser Part. Beams* **1990**, *8*, 247–252. [CrossRef]
14. Parigger, C.G.; Helstern, C.M.; Gautum, G. Temporally and spatially resolved emission spectroscopy of cyanide, hydrogen, and carbon in laser-induced plasma. *Atoms* **2019**, *7*, 74. [CrossRef]
15. Dackman, M. Laser-Induced Breakdown Spectroscopy for Analysis of High-Density Methane-Oxygen Mixtures. Master's Thesis, University of Tennessee Knoxville, Knoxville, TN, USA, 2014.
16. Griem, H.R. *Spectral Line Broadening by Plasmas*; Academic Press: New York, NY, USA, 1974.
17. Surmick, D.M. Spectroscopic Imaging of Aluminum Containing Plasma. Ph.D. Thesis, University of Tennessee Knoxville, Knoxville, TN, USA, 2016.
18. O'Haver, T. Peakfit.m. Available online: https://www.mathworks.com/matlabcentral/fileexchange/23611-peakfit-m (accessed on 22 May 2019).
19. Parigger, C.G.; Hornkohl, J.O. *Quantum Mechanics of the Diatomic Molecule with Applications*; IOP Publishing: Bristol, UK, 2019.
20. Parigger, C.G.; Woods, A.C.; Surmick, D.M.; Gautam, G.; Witte, M.J.; Hornkohl, J.O. Computation of diaotmic molecular spectra for selected transitions of aluminum monoxide, cyanide, diatomic carbon, and titanium monoxide. *Spectrochim Acta Part B At. Spectrosc.* **2015**, *107*, 132–138. [CrossRef]
21. Nelder, J.A.; Mead, R. A simplex-method for function minimization. *Comput. J.* **1965**, *7*, 308–313. [CrossRef]
22. Pretzler, G. A new method for numerical Abel-inversion. *Z. Naturforsch. A* **1991**, *46*, 639–641. [CrossRef]
23. Pretzler, G.; Jäger, H.; Neger, T.; Philipp, H.; Woisetschläger, J. Comparison of different methods of Abel inversion using computer simulated and experimental side-on data. *Z. Naturforsch. A* **1992**, *47*, 955–970. [CrossRef]
24. Kandel, Y.P. An Experimental Study of Hydrogen Balmer Lines in Pulsed Laser Plasma. Ph.D. Thesis, Wesleyan University, Middletown, CT, USA, 2009.
25. Gornushkin, I.B.; Shabanov, S.V.; Panne, U. Abel inversion applied to a transient laser induced plasma: Implications from plasma modeling. *J. Anal. At. Spectrom.* **2011**, *26*, 1457–1465. [CrossRef]
26. Killer, C. Abel Inversion Algorithm. Available online: https://www.mathworks.com/matlabcentral/fileexchange/43639-abel-inversion-algorithm (accessed on 30 June 2019).
27. Parigger, C.G.; Gautam, G.; Surmick, D.M. Radial electron density measurements in laser-induced plasma from Abel inverted hydrogen Balmer beta line profiles. *Int. Rev. At. Mol. Phys.* **2015**, *6*, 43–57.
28. Gautam, G. On Laser-induced Plasma Containing Hydrogen. Ph.D. Thesis, University of Tennessee Knoxville, Knoxville, TN, USA, 2017.
29. Parigger, C.G.; Surmick, D.M.; Gautam, G. Self-absorption characteristics of measured laser-induced plasma line shapes. *J. Phys. Conf. Ser.* **2017**, *810*, 012012. [CrossRef]
30. Gautam, G.; Parigger, C.G.; Helstern, C.M.; Drake, K.A. Emission spectroscopy of expanding laser-induced gaseous hydrogen-nitrogen plasma. *Appl. Opt.* **2017**, *56*, 9277–9284. [CrossRef] [PubMed]
31. Dardis, J.; Costello, J.T. Stagnation layers at the collision front between two laser-induced plasmas: A study using time-resolved imaging and spectroscopy. *Spectrochim. Acta Part B At. Spectrosc.* **2010**, *65*, 535–627. [CrossRef]
32. Zel'dovich, Y.B.; Raizer, Y.P. *Physics of Shock Waves and High-Temperature Hydrodynamic Phenomena*; Hayes, W.D., Probstein, R.F., Eds.; Academic Press: New York, NY, USA, 1966; Volume 1, pp. 347–348.

Publisher's Note: MDPI stays neutral with regard to jurisdictional claims in published maps and institutional affiliations.

Communication

Chirped Laser Pulse Effect on a Quantum Linear Oscillator

Valeriy A. Astapenko * and Evgeniya V. Sakhno

Moscow Institute of Physics and Technology (National Research University), Institutskij Per. 9, 141700 Dolgoprudnyj, Moscow Region, Russia; sakhno@phystech.edu

* Correspondence: astval@mail.ru

Received: 22 June 2020; Accepted: 30 July 2020; Published: 4 August 2020

Abstract: We present a theoretical study of the excitation of a charged quantum linear oscillator by chirped laser pulse with the use of probability of the process throughout the pulse action. We focus on the case of the excitation of the oscillator from the ground state without relaxation. Calculations were made for an arbitrary value of the electric field strength by utilizing the exact expression for the excitation probability. The dependence of the excitation probability on the pulse parameters was analyzed both numerically and by using analytical formulas.

Keywords: quantum linear oscillator; chirped laser pulse; photoexcitation

1. Introduction

The rapid development of the technique for generating short laser pulses with given parameters, including a frequency chirp [1], necessitates the development of adequate methods for the theoretical description of photo-processes in the field of such pulses with prescribed parameters. Along with the amplitude, carrier frequency, and pulse duration, an important parameter is the frequency chirp of the pulse. In papers [2–6], the features of excitation of a two-level system by chirped laser pulses were investigated. In work [2], the dependence of the population of the upper level of the quantum system on the chirp was calculated numerically and analytically for various values of the pulse duration and field amplitude. In particular, it was shown that in a certain range of parameters, the populations of a two-level system can be effectively controlled by variation of the chirp. In article [3], an effective scheme for controlling the superposition state of a two-level system using an ultrashort chirped laser pulse was proposed. In work [4], a high-precision population transfer was studied in a two-level model using a chirped Gaussian pulse.

In paper [7], the excitation of a classical Morse oscillator by a laser pulse with a linear frequency chirp was studied numerically for various values of the electric field strength and pulse duration. In particular, it was shown that there is a strong dependence of the oscillator excitation energy on the magnitude of the chirp, especially for multicycle pulses.

Paper [8] was devoted to numerical investigation of H atom ionization by chirped laser pulse. It was shown that chirped pulse more effectively ionizes atoms than the pulse with zero chirp.

In our previous paper [9], we investigated in detail the excitation of a quantum oscillator by short laser pulses without chirp using exact expression for the excitation probability obtained in [10]. It was shown that excitation probability as a function of carrier frequency and pulse duration is strongly dependent on the electric field amplitude in the pulse. In particular, criteria were established for the appearance of additional maxima in the probability of excitation for two types of envelopes of the laser pulse.

This work is a generalization of papers [9,10] in the case of a laser pulse with a linear frequency chirp. The main attention is paid to the influence of the frequency chirp on the probability of excitation of a quantum oscillator for various values of the carrier frequency and pulse duration.

2. Results

We considered a linear quantum oscillator excited by a laser pulse from the ground state. We assumed that pulse duration τ was sufficiently short so the condition $\tau < 1/\gamma$ was fulfilled (γ is oscillator relaxation constant) and the relaxation of the oscillator could be neglected.

According to paper [11], the following expression is appropriate for the probability of oscillator excitation from the ground state during the entire time of the pulse action:

$$W_{n0} = \frac{\overset{\rightharpoonup n}{n}}{n!}\exp(-\overline{n}). \tag{1}$$

Here, $\overline{n}$ is the average number of energy quanta at own frequency absorbed by oscillator during excitation. It is equal to (for the oscillator without relaxation)

$$\overline{n} = \frac{q^2}{2m\hbar\omega_0}\left|E(\omega_0)\right|^2 \tag{2}$$

Here, q, m, and ω_0 are the charge, mass, and own frequency of oscillator. $E(\omega)$ is the Fourier transform of electric field strength in the laser pulse. Furthermore, we considered pulse with Gaussian envelope and the linear frequency chirp.

Fourier transform of electric field strength in the Gaussian pulse with the linear frequency chirp has the form [12]:

$$E(\omega') = \frac{\sqrt{2\pi}E_0\tau}{\sqrt[4]{1+\alpha^2}}\exp\left\{-\frac{\omega'^2+\omega^2+2i\alpha\omega'\omega}{\Delta\omega^2}\right\}\cos\left\{0.5arctg(\alpha) - \frac{\alpha\left(\omega'^2+\omega^2\right)-2i\omega'\omega}{\Delta\omega^2}\right\} \tag{3}$$

Here, E_0 is the field amplitude, ω and τ are the carrier frequency and duration of laser pulse, α is the dimensionless chirp, and $\Delta\omega$ is the spectral width of the pulse which is equal to

$$\Delta\omega = \frac{\sqrt{1+\alpha^2}}{\sqrt{2}\tau}. \tag{4}$$

In the resonance approximation $|\omega - \omega_0| << \omega_0$ one has

$$\left|E(\omega,\tau,E_0,\alpha)\right|^2 \cong \frac{\pi}{2}\frac{E_0^2\tau^2}{\sqrt{1+\alpha^2}}\exp\left\{-\frac{(\omega_0-\omega)^2\tau^2}{1+\alpha^2}\right\}. \tag{5}$$

Let us introduce the following dimensionless parameters:

$$\zeta = \sqrt[4]{1+\alpha^2}\frac{\Omega_{10}}{\omega_0},\ \beta = \frac{\omega_0\tau}{\sqrt{1+\alpha^2}},\ \Delta = \frac{\omega-\omega_0}{\omega_0}, \tag{6}$$

where

$$\Omega_{10} = \frac{d_{10}E_0}{\hbar} = \frac{qE_0}{\hbar\sqrt{2m\omega_0}} \tag{7}$$

is the resonance Rabi frequency and d_{10} is the matrix element of the electric dipole moment for the $0 \to 1$ transition in the linear quantum oscillator. It is convenient for the analytical description of oscillator excitation to express the average number of absorbed quanta $\overline{n}$ via dimensionless parameters in the form

$$\overline{n}(\Delta,\beta,\zeta) = \frac{\pi}{2}\zeta^2\beta^2\exp\left(-\beta^2\Delta^2\right). \tag{8}$$

Here, we used Formulas (2) and (5)–(7).

Substituting Equation (8) in Equation (1), we obtained the formula for the numerical and analytical description of the excitation probability of the quantum linear oscillator by chirped laser pulse from the ground state.

The results of the numerical calculations are presented in the figures below for the excitation probability of transition $0 \to 1$ in the quantum oscillator for weak and strong fields and various values of the dimensionless frequency chirp. Calculations were made with the use of oscillator parameters q, m, and ω_0 corresponding to the vibration of the CO molecule in harmonic approximation.

Let us consider analytically the spectral dependence of the excitation probability of the transitions $0 \to n$ in the quantum oscillator. It was easy to obtain the position of spectral maxima using Formulas (1) and (8). In a weak field regime when

$$\Omega_{10}\tau < \sqrt{\frac{2n}{\pi}}\sqrt[4]{1+\alpha^2} \tag{9}$$

there is only one maximum at $\Delta = 0$ (see Figure 1a). With increasing electric field strength (i.e., Rabi frequency Ω_{10}), this maximum became a minimum. When the inverse to (9) inequality held, two maxima appeared at the following detunings of the carrier frequency from the own oscillator frequency (according to Figure 1b):

$$|\Delta_{1,2}| = \frac{\sqrt{1+\alpha^2}}{\omega_0\tau}\sqrt{\ln\left(\frac{\pi}{2n}\frac{\Omega_{10}^2\tau^2}{\sqrt{1+\alpha^2}}\right)} \tag{10}$$

One can see from this formula that the spectral distance between maxima in a strong field regime grew with the increase in chirp modulus and amplitude of the field.

The central spectral maximum turned into a minimum with the increasing field amplitude due to the depopulation of the ground state under the action of a laser pulse with a carrier frequency equal to the own frequency of the oscillator. The appearance of two maxima at a qualitative level can be associated with the emergence of quasienergy states under the action of a laser field.

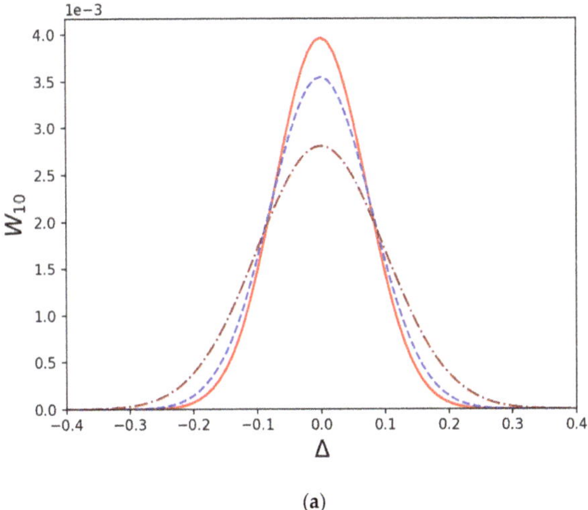

(a)

Figure 1. *Cont.*

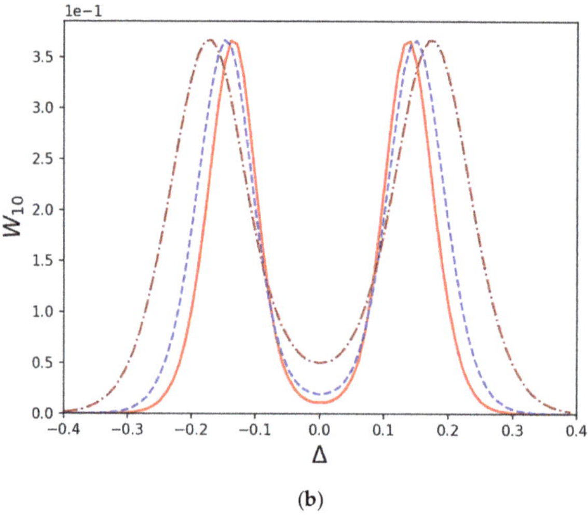

(**b**)

Figure 1. Spectrum of the excitation probability of transition $0 \to 1$ in quantum oscillator for weak field —$E_0 = 10^{-3}$ a.u. (**a**) strong field—$E_0 = 0.04$ a.u. (**b**) and different values of the dimensionless frequency chirp: solid line $\alpha = 0$, dotted line—$\alpha = 0.5$, dashed line—$\alpha = 1$.

Figure 2 demonstrates the dependence of the oscillator excitation probability at transition $0 \to 1$ as a function of dimensionless pulse duration (parameter β) for a weak (a) and strong (b) field and for different values of the frequency chirp.

Figure 2 shows that as the field amplitude increased, the maximum that was at $\beta = (100–150)$ in Figure 2a disappeared and became the minimum. Two new maxima appeared: one at $\beta << 100$ and the other at $\beta > 200$. The distance between the two maxima increased with the increasing magnitude of the chirp and of the field amplitude.

For the weak field amplitude, when the following inequality held (here, *e* is the base of the natural logarithm)

$$\Omega_{10} < \sqrt{\frac{2ne}{\pi} \frac{|\omega - \omega_0|}{\sqrt[4]{1+\alpha^2}}} \tag{11}$$

we had

$$\tau_{\max} = \frac{\sqrt{1+\alpha^2}}{|\omega - \omega_0|}. \tag{12}$$

For strong fields, when inverse to (11) inequality holding only an approximate analytical description of these maxima was possible. Then, one could obtain the following relations:

$$\tau_{\max,1} \cong \sqrt{\frac{2n}{\pi} \frac{\sqrt[4]{1+\alpha^2}}{\Omega_{10}}}, \ \tau_{\max,2} \cong \frac{\sqrt{1+\alpha^2}}{|\omega - \omega_0|} \sqrt{2.8 \ln\left(\sqrt{\frac{\pi}{2n}} \frac{\sqrt[4]{1+\alpha^2}\Omega_{10}}{|\omega - \omega_0|} \right)} \tag{13}$$

The resonance case ($\Delta = 0$) should be treated separately and the result was:

$$\tau_{\max} = \sqrt{\frac{2n}{\pi} \frac{\sqrt[4]{1+\alpha^2}}{\Omega_{10}}} \tag{14}$$

In conclusion, we gave an expression for the electric field strength amplitude corresponding to the maximum probability of the excitation of transition $0 \rightarrow n$ with other fixed parameters:

$$E_{0\max} = \frac{\hbar\omega_0}{d_{10}}\sqrt{\frac{2n}{\pi}\frac{\sqrt{1+\alpha^2}}{\omega_0\tau}}\exp\left\{\frac{(\omega-\omega_0)^2\tau^2}{2(1+\alpha^2)}\right\}. \tag{15}$$

Note that the effects considered in this paper in a strong field are due to the nonlinear nature of the interaction of the laser pulse with the quantum oscillator. The presence of chirp only modifies their manifestation.

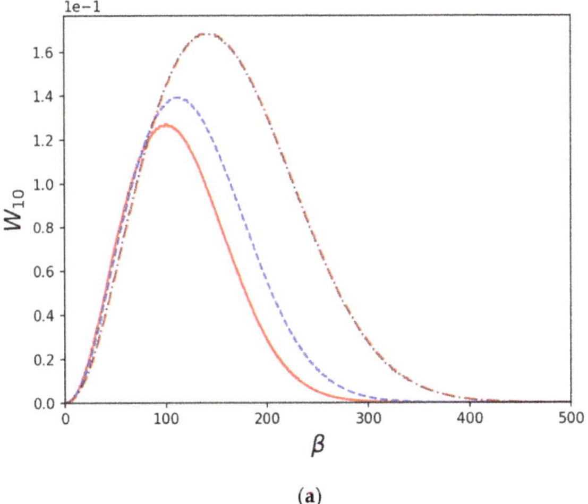

(**a**)

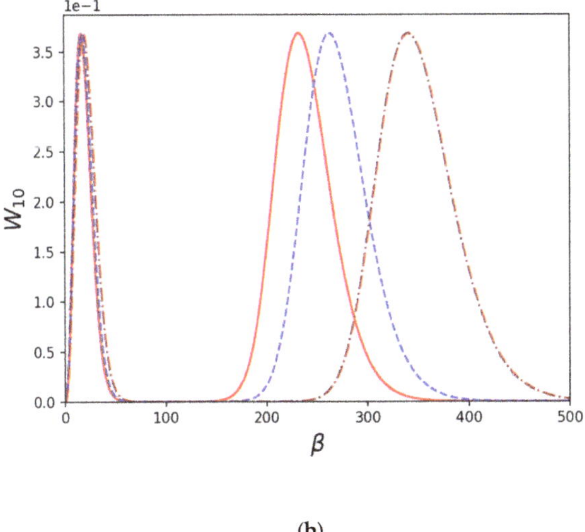

(**b**)

Figure 2. Excitation probability of transition $0 \rightarrow 1$ in the quantum oscillator as function of dimensionless pulse duration (β) for a weak field—$E_0 - 10^{-3}$ a.u. (**a**) strong field—$E_0 - 0.04$ a.u. (**b**) and different values of dimensionless chirp: solid line—α = 0, dotted line—α = 0.5, dashed line—α = 1.

3. Conclusions

Using the exact formula for the probability of exciting a quantum linear oscillator, we investigated the dependence of this probability on the carrier frequency and the pulse duration of laser pulse with Gaussian envelope at different values of the frequency chirp. Analytical expressions were derived that describe the features of oscillator excitation for different magnitudes of pulse parameters including frequency chirp. It was shown, in particular, that for weak fields the probability of excitation has one maximum as a function of the carrier frequency and pulse duration. With increasing electric field strength, a second maximum appears, the position of which depends on the frequency chirp value. With an increase in the magnitude of the chirp, these maxima shift to the region of large values of frequency detuning and pulse duration. In this case, the width of the maxima increases. Thus, by changing the magnitude of the chirp, one can control the probability of excitation of the quantum oscillator in a desired way.

Author Contributions: Both authors contributed equally. All authors have read and agreed to the published version of the manuscript.

Funding: This research received no external funding.

Acknowledgments: The research was supported by Moscow Institute of Physics and Technology in the framework of the 5-top-100 program (Project No. 075-02-2019-967).

Conflicts of Interest: The authors declare no conflict of interest.

References

1. Mitrofanov, A.V.; Sidorov-Biryukov, D.A.; Glek, P.B.; Rozhko, M.V.; Stepanov, E.A.; Shutov, A.D.; Ryabchuk, S.V.; Voronin, A.A.; Fedotov, A.B.; Zheltikov, A.M. Chirp-controlled high-harmonic and attosecond-pulse generation via coherent-wake plasma emission driven by mid-infrared laser pulses. *Opt. Lett.* **2020**, *45*, 750–753. [CrossRef] [PubMed]
2. Astapenko, V.A.; Romadanovskii, M.S. Excitation of a two-level system by a chirped laser pulse. *Laser Phys.* **2009**, *19*, 969–973. [CrossRef]
3. Jo, H.; Lee, H.G.; Guerin, S.; Ahn, J. Robust two-level system control by a detuned and chirped laser pulse. *Phys. Rev. A* **2017**, *96*, 033403. [CrossRef]
4. Dou, F.Q.; Liu, J.; Fu, L.B. High-fidelity superadiabatic population transfer of a two-level system with a linearly chirped Gaussian pulse. *Europhys. Lett.* **2017**, *116*, 60014. [CrossRef]
5. Ghaedi, Z.; Hosseini, M.; Sarreshtedari, F. Population change in the fine structure levels of cesium atoms using chirped laser. *J. Optoelectron. Nanostructures* **2017**, *2*, 1.
6. Ghaedi, Z.; Hosseini, M.; Sarreshtedari, F. Population engineering of the cesium atom fine structure levels using linearly chirped Gaussian laser pulse. *Opt. Commun.* **2019**, *431*, 109–114. [CrossRef]
7. Astapenko, V.A.; Romadanovskii, M.S. Excitation of the Morse oscillator by an ultrashort chirped pulse. *JETP* **2010**, *110*, 376–382. [CrossRef]
8. Prasad, V.; Dahiya, B.; Yamashita, K. Ionization of the H atom in ultrashort chirped laser pulses. *Phys. Scr.* **2010**, *82*, 055302. [CrossRef]
9. Astapenko, V.A.; Sakhno, E.V. Excitation of the 0→1 transition in a quantum oscillator under the action of an electromagnetic pulse of the Gaussian form. *Proc. MIPT* **2019**, *11*, 108–115. (In Russian)
10. Astapenko, V.A.; Sakhno, E.V. Excitation of a quantum oscillator by short laser pulses. *Appl. Phys. B* **2020**, *126*, 23. [CrossRef]
11. Husimi, K. Miscellanea in elementary quantum mechanics, II. *Prog. Theor. Phys.* **1953**, *9*, 381–402. [CrossRef]
12. Astapenko, V. *Interaction of Ultrafast Electromagnetic Pulses with Matter*; Springer Briefs in Physics: Heidelberg, Germany; New York, NY, USA; Dordrecht, The Netherland; London, UK, 2013; p. 94.

Article

Oscillatory-Precessional Motion of a Rydberg Electron Around a Polar Molecule

Eugene Oks

Physics Department, 380 Duncan Drive, Auburn University, Auburn, AL 36849, USA; oks@physics.auburn.edu

Received: 3 July 2020; Accepted: 10 July 2020; Published: 2 August 2020

Abstract: We provide a detailed classical description of the oscillatory-precessional motion of an electron in the field of an electric dipole. Specifically, we demonstrate that in the general case of the oscillatory-precessional motion of the electron (the oscillations being in the meridional direction (θ-direction) and the precession being along parallels of latitude (φ-direction)), both the θ-oscillations and the φ-precessions can actually occur on the same time scale—contrary to the statement from the work by another author. We obtain the dependence of φ on θ, the time evolution of the dynamical variable θ, the period T_θ of the θ-oscillations, and the change of the angular variable φ during one half-period of the θ-motion—all in the forms of one-fold integrals in the general case and illustrated it pictorially. We also produce the corresponding explicit analytical expressions for relatively small values of the projection p_φ of the angular momentum on the axis of the electric dipole. We also derive a general condition for this conditionally-periodic motion to become periodic (the trajectory of the electron would become a closed curve) and then provide examples of the values of p_φ for this to happen. Besides, for the particular case of $p_\varphi = 0$ we produce an explicit analytical result for the dependence of the time t on θ. For the opposite particular case, where p_φ is equal to its maximum possible value (consistent with the bound motion), we derive an explicit analytical result for the period of the revolution of the electron along the parallel of latitude.

Keywords: polar molecule; Rydberg electron; classical motion; periodic orbits

---

1. Introduction

An electron in the field of an electric dipole is the second most fundamental problem in atomic/molecular physics after the hydrogen atom—especially if the distance of the electron from the dipole is much larger than the dimension of the dipole, so that the dipole can be considered point-like. Theoretical studies of this fundamental system started as early as in 1947, when Fermi and Teller [1] investigated this problem in relation to the situation where a muon is slowly moving through a hydrogen atom. In the intervening years, many works have been published on this subject—see, e.g., paper [2] and references therein.

As for the classical motion, which is appropriate for a Rydberg electron in the field of an electric dipole, there were the following two theoretical studies for the case of the finite electric dipole (to the best of our knowledge). In 1968, Turner and Fox [3] considered the problem analytically in the elliptical coordinates (ξ, η, φ). They showed that for any finite size of the dipole, the bound motion exists in some region ($\xi_{min} < \xi < \xi_{max}$, $\eta_{min} < \eta < \eta_{max}$). The motion occurs inside a torus created by the revolution (about the dipole axis) of the two limiting ellipses, corresponding to $\xi = \xi_{min}$ and $\xi = \xi_{max}$, and of the two limiting parabolas, corresponding to $\eta = \eta_{min}$ and $\eta = \eta_{max}$.

In 2013, Kryukov and Oks [4] started by analytically considering circular orbits of a negative charge in the field of a finite dipole, when the orbital plane was perpendicular to the dipole axis; they used the cylindrical coordinates. (We note that as the application, the authors of paper [4] chose the finite dipole to be made by stationary proton and electron, while the negative charge moving in the field of this dipole was a muon; however, their analytical results are totally applicable to the motion of an electron in the field of any finite dipole.) They showed that stable circular orbits exist if the size of the dipole is greater or equal to some finite value (this value turned out to be the same as in many quantal studies of this system—see, e.g., paper [2] and references therein). Further, the authors of paper [4] went beyond the circular orbits and presented analytically a stable conic-helical motion of the electron in the field of the finite dipole: the helical orbit is confined to the surface of a right frustum of a cone coaxial with the dipole.

It is worth mentioning that obtaining the analytical results in papers [3,4] was possible because the general problem of the motion of a charged particle in the field of two stationary Coulomb centers (the problem, of which the motion of a charge in the field of a finite dipole is a particular case) is characterized by higher than geometrical (i.e., algebraic) symmetry. The symmetry is manifested by the presence of an additional conserved quantity: the projection of the super-generalized Runge–Lenz vector on the axis connecting the stationary charges [5].

As for the classical motion of a charged particle in the field of a point-like electric dipole, there were the following three theoretical studies (to the best of our knowledge). In 1968, Fox [6] treated this problem analytically and showed that the bound motion is possible only for the energy $E = 0$ and it is confined to a sphere. The author of paper [6] obtained the results in the form of quadratures, i.e., expressed them in terms of integrals, but without performing the integration and without a qualitative description of the motion (some details from paper [6] are reproduced in the next section of the present paper).

In 1995, Jones [7] derived analytical results for a semicircular orbit along a meridian on a sphere, to which the motion is confined (confined according to Fox result [6]). He considered the moving particle to be positively charged and pointed out that its motion is identical to the motion of a pendulum: the potential of a simple pendulum is identical to the potential of the dipole at a constant radius.

In 1996, McDonald [8] focused at two types of circular (or semicircular) orbits. One type was the same as that presented by Jones [7]. Another type was a circular orbit along a certain parallel of latitude. For the general case, McDonald [8] reproduced some analytical results from Fox's paper [6] (without referring to that paper) and then mentioned that in general the motion should consist of large oscillations with respect to the polar angle θ combined with a slow precession about the z-axis (the z-axis being the dipole axis).

In the present paper we start where Fox stopped in paper [6]. We provide a detailed classical description of the oscillatory-precessional motion of an electron in the field of an electric dipole. Specifically, we demonstrate that in the general case of the oscillatory-precessional motion of the electron (the oscillations being in the meridional direction (θ-direction) and the precession being along parallels of latitude (φ-direction)), both the θ-oscillations and the φ-precessions can actually occur on the same time scale, so the statement to the contrary from work [8] is incorrect.

We obtain the dependence of φ on θ in the form of a one-fold integral in the general case and illustrate it pictorially. We also derive an explicit analytical result for $\varphi(\theta)$ for relatively small values of the projection p_φ of the angular momentum on the axis of the electric dipole.

For the particular case of $p_\varphi = 0$ we produce an explicit analytical result for the dependence of the time t on θ. For the opposite particular case, where p_φ is equal to its maximum possible value (consistent with the bound motion), we derive an explicit analytical result for the period of the revolution of the electron along the parallel of latitude.

We obtain the time evolution of the dynamical variable θ, the period T_θ of the θ-oscillations, and the change of the angular variable φ during one half-period of the θ-motion—all in the form of one-fold integrals in the general case and illustrate it pictorially. We also produce the corresponding explicit analytical expressions for relatively small values of the projection p_φ of the angular momentum on the axis of the electric dipole.

Finally, we derive a general condition for this conditionally-periodic motion to become periodic (the trajectory of the electron would become a closed curve) and then provide examples of the values of K for this to happen.

2. Classical Non-Circular Orbits

Following Fox [6], we consider an electric dipole with the dipole moment D centered at the origin. The motion of an electron in the field of the dipole is analyzed in spherical polar coordinates (r, θ, φ) with the z-axis chosen along the dipole axis, such that the positive pole points to the upper hemisphere. In this reference frame the energy has the following form:

$$E = m[(dr/dt)^2 + r^2(d\theta/dt)^2 + r^2\sin^2\theta\,(d\varphi/dt)^2]/2 - eD\cos\theta/r^2, \quad (1)$$

where m and e are the mass and the absolute value of the electron charge.

Due to the axial symmetry, the z-projection of the electron angular momentum M_z is conserved, as manifested by the fact that the energy, while depending on dφ/dt, does not depend on φ. Fox [6] denoted M_z as p_φ because it is the generalized momentum corresponding to the dynamical variable φ:

$$p_\varphi = mr^2\sin^2\theta\,(d\varphi/dt) = \text{const}. \quad (2)$$

Due to the above symmetry, the θ-motion and the φ-motion can be separated.

Fox [6] showed that the bound motion is possible only for E = 0 and r = const. Thus, the bound motion is confined to a sphere and the dynamical variables are only θ and φ.

For the θ-motion, Fox [6] derived the following differential equation:

$$(dx/dt)^2 = [2eD/(mr^4)](-x^3 + x - K), \quad (3)$$

where

$$x = \cos\theta, \qquad K = p_\varphi{}^2/(2meD). \quad (4)$$

The corresponding equation for the φ-motion follows from Equation (2):

$$d\varphi/dt = p_\varphi/(mr^2\sin^2\theta) = p_\varphi/[mr^2(1 - x^2)]. \quad (5)$$

After finding x(t) from Equation (3), one can obtain φ(t) from Equation (5).

Fox [6] noted that for the bound motion to occur, the polynomial

$$y(x) = -x^3 + x - K \quad (6)$$

in Equation (3) must be positive in some range of x within the interval from −1 to 1. This polynomial has a negative minimum equal to $-K - 2/3^{3/2}$ at $x = -1/3^{1/2}$ and a maximum equal to $-K + 2/3^{3/2}$ at $x = 1/3^{1/2}$. Obviously, for the range of the bound motion to exist, the maximum should be positive, leading to the following requirement (Fox [6]):

$$K < 2/3^{3/2} = K_{max} \quad (7)$$

or equivalently (see Equation (4))

$$D > 3^{3/2}p_\varphi{}^2/(4me). \quad (8)$$

Under the condition (7) the motion with respect to the dynamical variable x can be confined between two positive turning points, so that the bound motion occurs in the upper hemisphere, as noted by Fox [6]. We note in passing that, according to Equation (8), classical bound states exist for any value of the dipole moment D—because the generalized momentum p_φ in the right side of Equation (8) can be zero. The same is true for classical bound states for any finite size of the dipole, as shown in paper [3]. In contrast, the quantum bound states exist only if the dipole moment exceeds some critical value D_{min} = 0.6393148771999813 (in atomic units). This value of D_{min} with the accuracy of the first 16 digits was calculated in paper [2] of 2007 (paper [2] contains references to most of the previous calculations of D_{min}). However, the existence of the critical dipole moment D_{min} and its first three digits were calculated as early as in 1947 by Fermi and Teller [1].

From this point on, we present new results. Figure 1 shows a three-dimensional plot of the polynomial y(x, K) as x varies from −1 to 1 and K varies from 0 to $K_{max} = 2/3^{3/2}$. It is seen that in this range of K, the maximum value of y is positive, so that there is indeed a range of x that allows classical bound motion.

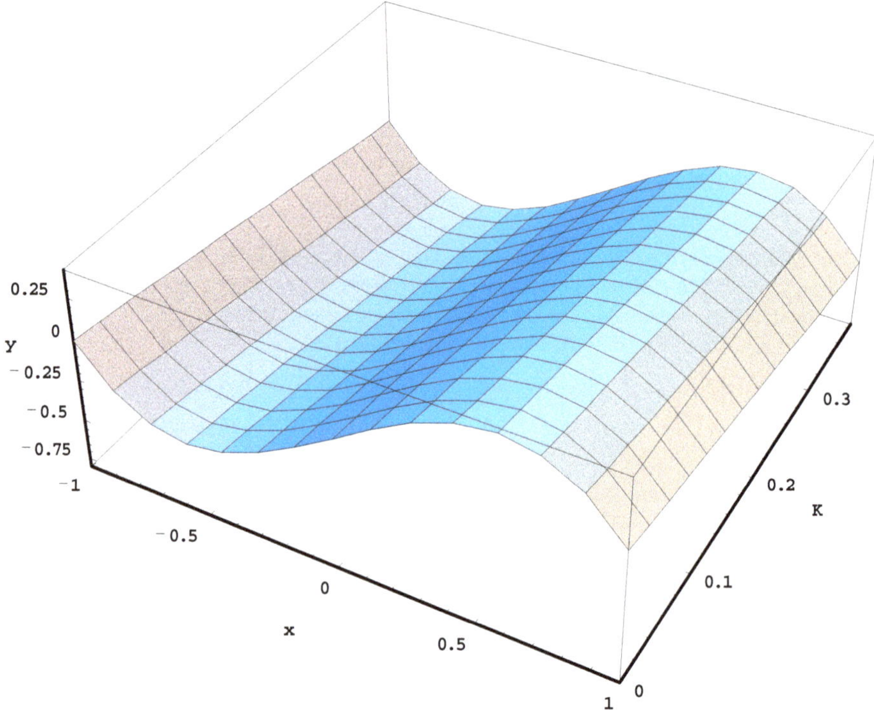

Figure 1. Three-dimensional plot of the polynomial y(x, K) from Equation (6).

We denote positive turning points as x_2 and x_3 ($x_3 > x_2$. They are the real roots of the following cubic equation:

$$x^3 - x + K = 0. \tag{9}$$

By solving Equation (9) we obtain the following explicit expressions for the turning points:

$$x_2(K) = (-1)^{2/3}2^{1/3}/[(729K^2 - 108)^{1/2} - 27K]^{1/3} - (-1)^{1/3}[(729K^2 - 108)^{1/2} - 27K]^{1/3}/(2^{1/3}3). \tag{10}$$

$$x_3(K) = 2^{1/3}/[(729K2 - 108)^{1/2} - 27K]^{1/3} - [(729K2 - 108)^{1/2} - 27K]^{1/3}/(2^{1/3}3). \tag{11}$$

Figure 2 shows a plot of both $x_2(K)$ and $x_3(K)$, where $x_2(K)$ and $x_3(K)$ are the lower part and the upper part of the double-valued curve, respectively. The lower and upper parts intersect at $K = K_{max} = 2/3^{3/2}$, where $x_2(2/3^{3/2})$ and $x_3(2/3^{3/2}) = 1/3^{1/2}$.

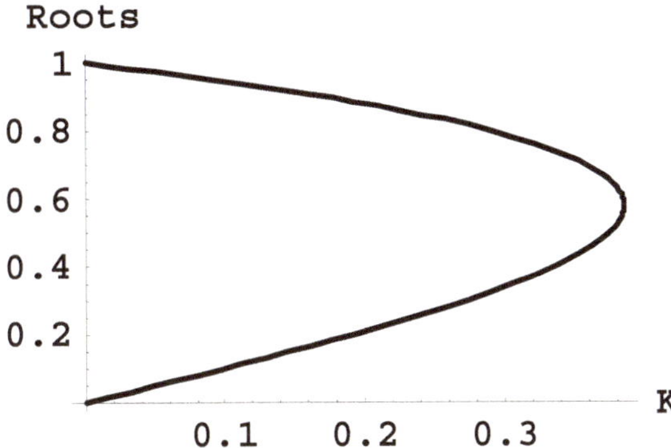

Figure 2. Plot of both positive roots x_2 and x_3 of the cubic Equation (8) versus $K = p_\varphi^2/(2meD)$.

$x_2(K)$ and $x_3(K)$ are the lower part and the upper part of the double-valued curve, respectively. We introduce a scaled dimensionless time τ as follows:

$$\tau = t\,[2\,eD/(mr^4)]^{1/2}. \tag{12}$$

Then Equation (3) can be represented in the form

$$d\tau = \pm dx/(-\,x^3 + x - K)^{1/2}. \tag{13}$$

(We note that Fox [6] chose only the minus sign in his Equation (12) analogous to our Equation (13); below we show that both signs should be considered.)

Now we study limiting cases. In the special case of $K = K_{max} = 2/3^{3/2}$, there is no θ-motion: the electron follows a circular path along the parallel of latitude corresponding to $\cos\theta = 1/3^{1/2}$. The latter equation yields $\theta = 0.9553$ rad $= 54.74$ degrees. (We note that McDonald [8] considered this circular motion for a positive charge, in which case $\cos\theta = -1/3^{1/2}$, so that $\theta = 2.1863$ rad $= 125.26$ degrees.) From Equation (5) it follows that the electron rotates with the constant angular velocity

$$d\varphi/dt = 3p_\varphi/(2mr^2), \tag{14}$$

corresponding to the period

$$T = 4\pi mr^2/(3p_\varphi). \tag{15}$$

From Equation (4) it follows that $p_\varphi = (2KmeD)^{1/2}$, so that for $K = K_{max} = 2/3^{3/2}$ we have $p_\varphi = 2(meD)^{1/2}/3^{3/4}$, so that Equations (14) and (15) can be rewritten as follows:

$$d\varphi/dt = 3^{1/4}[eD/(mr^4]^{1/2},\ T = (2\pi/3^{1/4})\,[mr^4/(eD)]^{1/2}. \tag{16}$$

In the opposite limit of $K << 1$, Equations (9) and (10) simplify to:

$$x_2(K) \approx K,\ x_3(K) \approx 1 - K/2. \tag{17}$$

In the special case of $K = 0$, i.e., $p_\varphi = 0$, there is no φ-motion. The electron oscillates along a semicircle located in a meridional plane in the upper hemisphere. (This is analogous to the corresponding result by Jones [7], later reproduced by McDonald [8], for a positive charge: the only difference is that for the positive charge, the semicircular orbit is in the lower hemisphere.) Let us study this special case in more detail before proceeding to the general case.

For K = 0, Equation (13) can be integrated analytically to yield the following explicit dependence of the scaled time τ on x, i.e., the dependence of τ on $\cos\theta$:

$$\tau = -\{\pm 2i\,\mathbf{F}[\arcsin(-x)^{1/2}, -1]\}, \tag{18}$$

where $\mathbf{F}(\alpha, q)$ is the incomplete elliptic integral of the first kind. Despite the formal appearance of the imaginary unit i in Equation (18), the right side of Equation (18) is actually real for the range of x from 0 to 1 where the motion occurs. In particular, for $x \ll 1$, we obtain from Equation (18) the following:

$$x \approx \pm\tau/2. \tag{19}$$

Figure 3 shows the plot of both branches of the dependence $\tau(x)$ from Equation (18): the upper and lower parts of the double-valued curve corresponds to the two different signs in the right side of Equation (18). The zero value of τ corresponds to x = 0, i.e., to $\theta = \pi/2$, when the electron is at the equator of the upper hemisphere. We note that $\tau(\pm 1) = \pm\ 2.622$. If we would start following the motion at $\tau(-1) = -2.622$, i.e., when the electron is at the north pole, we would follow the lower branch of the double-valued curve in Figure 3 from x = 1 (the electron at the equator) to x = 0 at $\tau = 0$ (the electron at the north pole), then switch to the upper branch and follow it from x = 0 to x = 1 at $\tau(1) = 2.622$, when the electron is again at the equator. It should be emphasized that at x = 0, the geographical longitude of the electron in the upper hemisphere changes by 180 degrees.

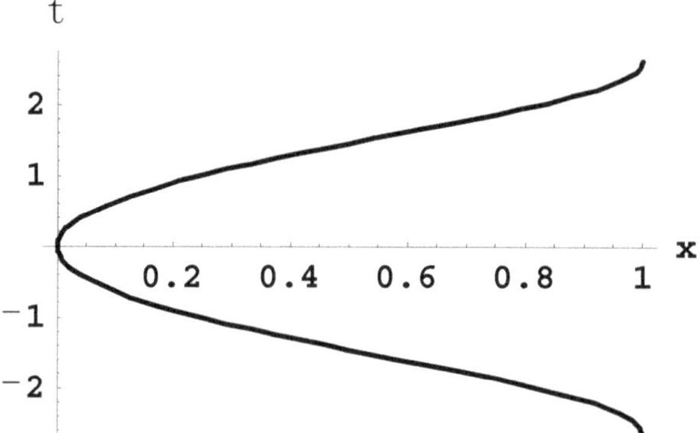

Figure 3. Dependence of the scaled time τ (defined in Equation (12)) on $x = \cos\theta$ during one half-period of the electron oscillation along a semicircular path through the north pole of the upper hemisphere (K = 0).

The temporal and spatial evolutions of the electron depicted in Figure 3 obviously represent one half-cycle of its oscillations. The full period of oscillations is as follows (in terms of the scaled time):

$$\tau_0 = 4\tau(1) = 10.488. \tag{20}$$

In the usual units this corresponds to the period

$$T = 10.488\ [mr^4/(2eD)]^{1/2}. \tag{21}$$

Now we come back to the general case of an arbitrary K from the interval (0, $K_{max} = 2/3^{3/2}$). Based on Equation (13), the dependence of the scaled time τ on x (i.e., the dependence of τ on $\cos\theta$) in the general case:

$$\tau(x, K) = \pm \int_{x2(K)}^{x} dz/\left(-z^3 + z - k\right)^{1/2} \tag{22}$$

where the lower limit of the integration is the smaller of the two turning points. Figure 4 shows the evolution of the scaled time τ during one period of the θ-motion—i.e., as x varies from $x_2(K)$ to $x_3(K)$ and back to $x_2(K)$—for K = 0.1 (solid line), K = 0.2 (dashed line), and K = 0.3 (dotted line).

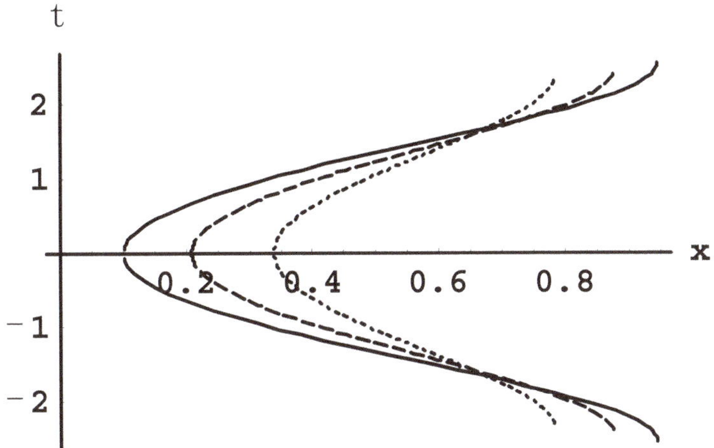

Figure 4. Evolution of the scaled time τ during one period of the θ-motion for K = 0.1 (solid line), K = 0.2 (dashed line), and K = 0.3 (dotted line).

For K << 1, the integral in Equation (22) can be calculated analytically. Namely, we expand this integral in Taylor series up to (including) the terms ~ K and then calculate the emerging integrals analytically to obtain the following:

$$\tau(x, K) \approx \pm\{f(x) - f[x_2(K)] + Kg(x) - Kg[x_2(K)]\}. \tag{23}$$

Here

$$f(x) = -2i\,\mathbf{F}[\arcsin(-x)^{1/2}, -1], \tag{24}$$

$$g(x) = [3x^2 - 2 - x^2(1 - x^2)^{1/2}\,{}_2F_1(3/4, 1/2, 7/4, x^2)]/[2(x - x^3)^{1/2}] \tag{25}$$

where ${}_2F_1(a, b, c, z)$ is the hypergeometric function.

Figure 5 illustrates the accuracy of the approximate analytical result for $\tau(x, K)$ from Equation (23) for K = 0.01 (solid line) by the comparison with the corresponding exact result obtained by the numerical integration in Equation (22) (dashed line). It can be seen that the accuracy of the analytical result is very good.

Now we calculate the scaled period T_θ of the θ-motion. It is calculated by the following formula (in units of $mr^4/(2eD)$):

$$T_\theta(K) = 2\int_{x2(K)}^{x3(K)} dz/\left(-z^3 + z - k\right)^{1/2} \tag{26}$$

The dependence of the scaled period T_θ on K is shown in Figure 6.

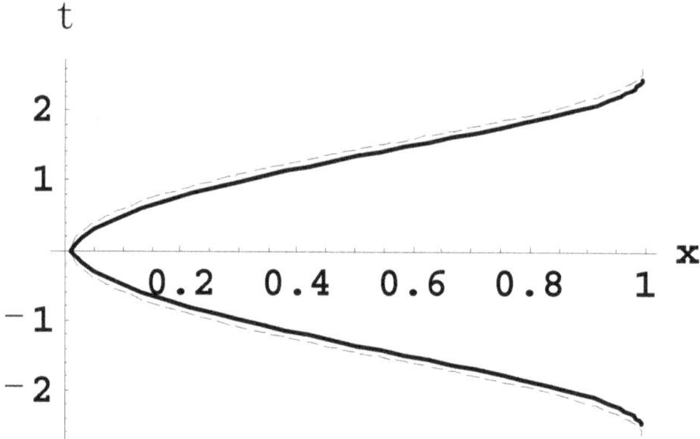

Figure 5. Comparison of the approximate analytical result for τ(x, K) from Equation (23) for K = 0.01 (solid line) with the corresponding exact result obtained by the numerical integration in Equation (22) (dashed line).

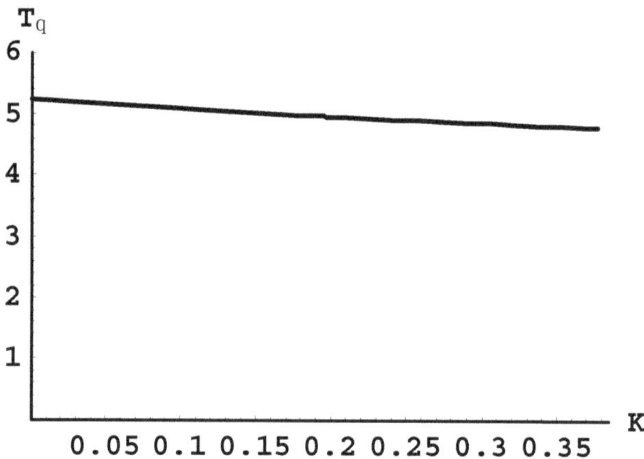

Figure 6. Dependence of the scaled period T_θ of the θ-motion on the parameter $K = p_\varphi{}^2/(2meD)$. The period T_θ is in units of $mr^4/(2eD)$.

For K << 1, an explicit analytical result for the scaled period of the θ-motion is as follows:

$$T_\theta(K) \approx 2\,\{f\,(1 - K/2) - f\,(K) + Kg(1 - K/2) - Kg(K)\}, \tag{27}$$

where functions f and g are defined by Equations (24) and (25), respectively.

Now we proceed to analyzing the φ-motion. Equation (5) can be rewritten in the form

$$d\varphi = p_\varphi dt/[mr^2(1 - x^2)]. \tag{28}$$

Then by using the relation between dt and dx from Equation (3) and applying the integration with respect to x, we obtain the following dependence of the angular variable φ and the angular variable x = cosθ:

$$\varphi(K,x) = K^{1/2}\int_{x2(K)}^{x} dz/\left[\left(1-Z^2\right)\left(-z^3+z-k\right)^{\frac{1}{2}}\right] \tag{29}$$

Figure 7 presents the dependence of φ on x during one half-period of the θ-motion (i.e., as x varies from $x_2(K)$ to $x_3(K)$) for K = 0.1 (solid line), K = 0.2 (dashed line), and K = 0.3 (dotted line). It is seen that as the parameter K increases, the curve $\varphi(x)$ becomes steeper and the change of φ over one half-period of the θ-motion slightly increases.

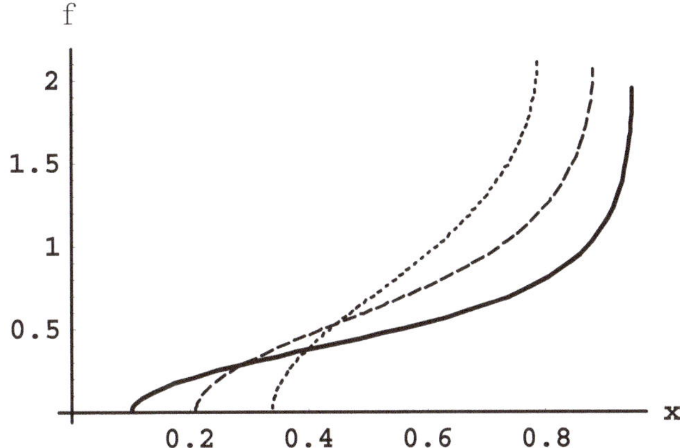

Figure 7. Dependence of φ on $x = \cos\theta$ during one half-period of the θ-motion for K = 0.1 (solid line), K = 0.2 (dashed line), and K = 0.3 (dotted line).

For K << 1, we obtain the following explicit analytical result for $\varphi(K, x)$

$$\varphi(K, x) \approx K^{1/2}[j(x) - j(K)] \tag{30}$$

where

$$j(x) = [x/(1 - x^2)]^{1/2} \{1 - (x - 1/x)^{1/2} \mathbf{F} [\text{arccsc}(x^{1/2}), -1]\} \tag{31}$$

Here $\mathbf{F}(\alpha, q)$ is the incomplete elliptic integral of the first kind. In Equation (30) we used the fact that $x_2(K) \approx K$ for K << 1.

Figure 8 illustrates the accuracy of the approximate analytical result for $\varphi(K, x)$ from Equation (30) for K = 0.02 (solid line) by the comparison with the corresponding exact result from Equation (29) (dashed line). It seen that the accuracy of the analytical result is very good.

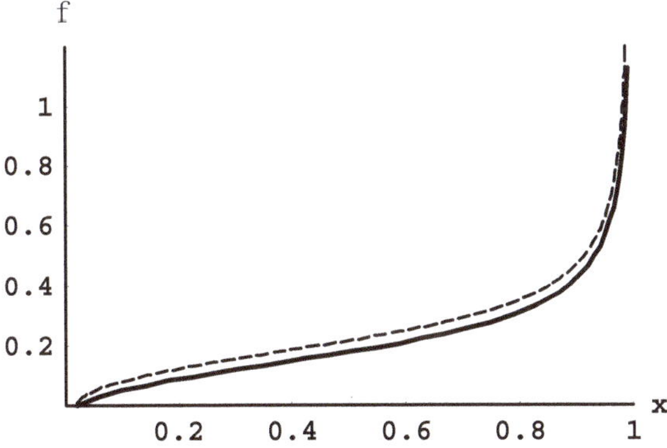

Figure 8. Comparison of the approximate analytical result for $\varphi(K, x)$ from Equation (30) for K = 0.02 (solid line) with the corresponding exact result from Equation (29) (dashed line).

The change of the angular variable φ during one half-period of the θ-motion is

$$\Delta\varphi(K) = K^{1/2} \int_{x2(K)}^{x3(K)} dz/\left[\left(1 - Z^2\right)\left(-z^3 + z - k\right)^{\frac{1}{2}}\right] \tag{32}$$

For K close to zero, we have $\Delta\varphi \approx \pi/2$, while for K close to $K_{max} = 2/3^{3/2}$, we have $\Delta\varphi \approx \pi/2^{1/2}$, so that

$$1/2 < \Delta\varphi/\pi < 1/2^{1/2}. \tag{33}$$

Between these two limits, Δφ monotonically increases as K grows.

The combination of the θ-motion and φ-motion exhibits the oscillatory-precessional behavior of the electron: the oscillations in the upper hemisphere in the meridional direction combined with the precession along parallels of latitude. Specifically, during one period $T_\theta(K)$ of the θ-oscillation (given by Equation (26) and presented in Figure 6 in units of $mr^4/(2eD)$), the angle φ advances by Δφ(K) given by Equation (32). In general, Δφ(K) is not equal to nπ/m, where n and m are relatively small integers, so that the combined motion is conditionally-periodic: the trajectory generally is not a closed curve.

However, in some particular cases, where

$$\Delta\varphi(K) = n\pi/m, \tag{34}$$

the trajectory becomes a close curve and the motion becomes periodic. Below we present three examples corresponding to the three lowest pairs of integers n and m in Equation (34): (n = 2, m = 3), (n = 3, m = 5), and (n = 4, m = 7). Figure 9 illustrates where (i.e., at which values of K) the curve Δφ(K)/π intersects n/m for the above three pairs.

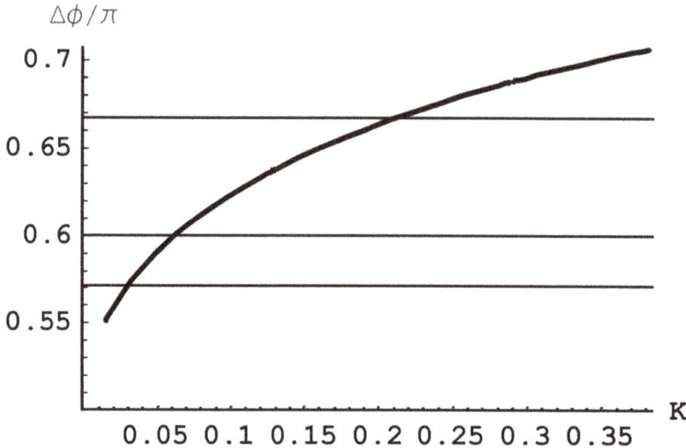

Figure 9. Intersections of the curve Δφ(K)/π (thick line) with the tree horizontal lines corresponding to 2/3 (the top thin line), 3/5 (the middle thin line), and 4/7 (the bottom thin line). For the values of K, corresponding to these intersections, the motion, which is generally conditionally-periodic, becomes truly periodic.

Here are the details for all three cases. In the case of n = 2, m = 3, during three periods of the θ-oscillations, the angle φ completes two full circles. This happens for K = 0.2103.

In the case of n = 3, m = 5, during five periods of the θ-oscillations, the angle φ completes three full circles. This happens for K = 0.0632.

In the case of n = 4, m = 7, during seven periods of the θ-oscillations, the angle φ completes four full circles. This happens for K = 0.0310.

These examples show, in particular, that the θ-oscillations can actually occur on the same time scale as the φ-precessions. The same is true in the general case of the conditionally-periodic orbits of the electron. Thus, the statement from work [8] that "in general the motion should consist of large oscillations with respect to the polar angle θ combined with a slow precession about the z-axis" is incorrect.

3. Conclusions

We considered the classical bound motion of a Rydberg electron around a polar molecule. We showed that in the general case, where the motion consists of the oscillations in the upper hemisphere in the meridional direction (θ-direction) combined with the precession along parallels of latitude (φ-direction), both the θ-oscillations and the φ-precessions can actually occur on the same time scale, so that the statement to the contrary from work [8] is incorrect. (We also corrected one of the equations in paper [6].)

We obtained the relation between the two dynamical variables, i.e., the dependence of φ on θ in the form of a one-fold integral in the general case and illustrated it pictorially. We also derived an explicit analytical result for φ(θ) in the case where the dimensionless parameter $K = p_\varphi^2/(2meD) << 1$, i.e., for relatively small values of the projection p_φ of the angular momentum on the axis of the electric dipole.

For the particular case of K = 0, where the electron oscillates along a semicircle crossing the north pole, we derived an explicit analytical result for the dependence of time t on θ. For the opposite particular case, where $K = K_{max} = 2/3^{3/2}$ and the electron follows a circular path along the parallel of latitude corresponding to θ = 0.9553 rad = 54.74 degrees, we obtained an explicit analytical result for the period of the revolution.

Further, we obtained the time evolution of the dynamical variable θ and the period T_θ of the θ-oscillations in the form of a one-fold integral in the general case and illustrated it pictorially. We also derived the corresponding explicit analytical expressions for the case of K << 1.

We also obtained the change of the angular variable φ during one half-period of the θ-motion in the form of a one-fold integral in the general case. We provided a pictorial illustration of this result.

Finally, we studied whether there are values of the parameter K, such that this conditionally-periodic motion would become truly periodic, so that the trajectory of the electron would become a closed curve. We derived a general condition for this to happen and then provided three examples of the values of K, enabling the motion to become periodic.

We believe that our classical results provide a physical insight into the complicated dynamics of a Rydberg electron around a polar molecule.

Funding: This research received no external funding.

Conflicts of Interest: The author declares no conflict of interest.

References

1. Fermi, E.; Teller, E. The capture of negative mesotrons in matter. *Phys. Rev.* **1947**, *72*, 399–408. [CrossRef]
2. Connolly, K.; Griffiths, D.J. Critical dipoles in one, two, and three dimensions. *Am. J. Phys.* **2007**, *75*, 524–531. [CrossRef]
3. Turner, J.E.; Fox, K. Classical motion of an electron in an electric-dipole field. I. Finite dipole case. *J. Phys. A* **1968**, *1*, 118–123. [CrossRef]
4. Kryukov, N.; Oks, E. Muonic–electronic negative hydrogen ion: Circular states. *Can. J. Phys.* **2013**, *91*, 715–721. [CrossRef]
5. Kryukov, N.; Oks, E. Super-generalized Runge-Lenz vector in the problem of two Coulomb or Newton centers. *Phys. Rev. A* **2012**, *85*, 054503. [CrossRef]
6. Fox, K. Classical motion of an electron in an electric-dipole field. II. Point dipole case. *J. Phys. A (Proc. Phys. Soc.)* **1968**, *1*, 124–127. [CrossRef]

7. Jones, R.S. Circular motion of a charged particle in an electric dipole field. *Am. J. Phys.* **1995**, *63*, 1042–1043. [CrossRef]
8. McDonald, K.T. Motion of a Point Charge near an Electric Dipole. 1996. Available online: http://www.hep.princeton.edu/~{}mcdonald/examples/dipole.pdf (accessed on 1 July 2020).

Article

Application of the Generalized Hamiltonian Dynamics to Spherical Harmonic Oscillators

Eugene Oks

Physics Department, Auburn University, 380 Duncan Drive, Auburn, AL 36849, USA; goks@physics.auburn.edu

Received: 9 June 2020; Accepted: 29 June 2020; Published: 7 July 2020

Abstract: Dirac's Generalized Hamiltonian Dynamics (GHD) is a purely classical formalism for systems having constraints: it incorporates the constraints into the Hamiltonian. Dirac designed the GHD specifically for applications to quantum field theory. In one of our previous papers, we redesigned Dirac's GHD for its applications to atomic and molecular physics by choosing integrals of the motion as the constraints. In that paper, after a general description of our formalism, we considered hydrogenic atoms as an example. We showed that this formalism leads to the existence of classical non-radiating (stationary) states and that there is an infinite number of such states—just as in the corresponding quantum solution. In the present paper, we extend the applications of the GHD to a charged Spherical Harmonic Oscillator (SHO). We demonstrate that, by using the higher-than-geometrical symmetry (i.e., the algebraic symmetry) of the SHO and the corresponding additional conserved quantities, it is possible to obtain the classical non-radiating (stationary) states of the SHO and that, generally speaking, there is an infinite number of such states of the SHO. Both the existence of the classical stationary states of the SHO and the infinite number of such states are consistent with the corresponding quantum results. We obtain these new results from first principles. Physically, the existence of the classical stationary states is the manifestation of a non-Einsteinian time dilation. Time dilates more and more as the energy of the system becomes closer and closer to the energy of the classical non-radiating state. We emphasize that the SHO and hydrogenic atoms are not the only microscopic systems that can be successfully treated by the GHD. All classical systems of N degrees of freedom have the algebraic symmetries O_{N+1} and SU_N, and this does not depend on the functional form of the Hamiltonian. In particular, all classical spherically symmetric potentials have algebraic symmetries, namely O_4 and SU_3; they possess an additional vector integral of the motion, while the quantal counterpart-operator does not exist. This offers possibilities that are absent in quantum mechanics.

Keywords: generalized Hamiltonian dynamics; spherical harmonic oscillator; classical non-radiating stationary states; algebraic symmetry of classical systems

---

1. Introduction

The generalized Hamiltonian dynamics (hereafter GHD) was developed by Dirac 70 years ago [1–3]. While the conventional Hamiltonian dynamics employs an assumption that the momenta are independent functions of velocities, Dirac considered a more general situation where momenta are not independent functions of velocities [1–3]. From the physical point of view, the GHD is a purely classical formalism for constrained systems: in the GHD, constraints are incorporated into the Hamiltonian. Dirac developed the GHD with the purpose to apply it to quantum field theory [3].

For the application to the quantum field theory and statistical mechanics, Dirac's GHD was further developed by a number of authors—see, e.g., papers by Sergi [4–7] and references therein. The focus of Sergi's works [4–7] was on non-Hamiltonian mathematical structures, including non-Hamiltonian commutators.

In search of a purely classical formalism that can be applied to atomic and molecular physics and can reproduce quantum results classically, Oks and Uzer, in 2002 [8], brought up an idea of choosing integrals of the motion as the constraints. The authors of paper [8] first provided a general description of their formalism. Then they considered hydrogenic atoms as an example. The authors of paper [8] demonstrated that this purely classical formalism allows the existence of classical non-radiating states, so that such states are stable. Remember that, according to the usual classical formalism (including classical electrodynamics), the electron would lose the energy through the radiation and fall into the nucleus: this failure of the usual classical formalism was one of the primary reasons for the birth of quantum mechanics. In distinction, in the purely classical formalism from paper [8] the electron does not fall into the nucleus.

Further, the authors of paper [8] derived the formula for the energy of such classical non-radiating states. They showed that this set of classical energies coincides with the energies of the corresponding quantal stationary states. While obtaining this result, the authors of paper [8] did not "forcefully" quantize any physical quantity describing the atom.

It should be emphasized that the physical interpretation was that the existence of these classical non-radiating states is due to a new kind of a time dilation. This new kind of the time dilation is non-Einsteinian (see paper [9] and book [10]): it has nothing to do with the time-dilation in the theory of relativity.

The purpose of the present paper is to present another application of the Oks–Uzer purely classical formalism [8] within atomic and molecular physics, i.e., to microscopic systems of discrete (rather than continuous) charges. Before proceeding with our presentation, we note in passing that, outside atomic and molecular physics, some authors studied whether there are continuous moving charge distributions that would not radiate—see, e.g., the paper by Goedecke [11] and references therein. It was found that certain continuous charge distributions would not radiate. However, first, the radius of their "orbit" should be less than the size of the charge distribution (so that the the distribution would only "wobble"). Second, this result is valid only for continuous charge distributions, so that atoms and molecules do not qualify.

In our view, the most interesting (and potentially relevant to atomic physics) paper of that series was published by Raju [12]. He considered classical circular orbits of the electron and of the proton in a hydrogen atom. He took into account the relativistic effect of the retardation, due to which the force on the electron is at the "last-seen" position of the proton, while the proton has moved since then. This results in a torque that would initially accelerate the electron and later on decelerate the electron and so on. Then Raju [12] added radiative damping into the consideration, which provides a decelerating torque for the electron. Raju [12] found sets of parameters for which the initially accelerating torque due to the retardation would be totally compensated by the decelerating torque due to the radiative damping. Then Raju [12] stated that "it was prematurely concluded that radiative damping makes the classical hydrogen atom unstable". However, this statement seems to be incorrect. In reality, within Raju's concept [12], the radiative damping does make the classical hydrogen atom unstable. Indeed, when the retardation torque compensates the radiative damping torque, this means only that the tangential acceleration of the electron vanishes, but the centripetal acceleration of the electron remains and so does the radiation. Another view of this situation is that the two torques can compensate each other, but one of them is due to the radiation, which carries the energy away from the electron. The electron would therefore continuously lose energy and would fall into the proton. Thus, the concept by Raju [12] did not lead to a non-radiating state of hydrogen atoms.

To avoid any confusion, we also mention that there were attempts to find stable states of hydrogen atoms in frames of a so-called stochastic electrodynamics, where the interaction with the zero-point fluctuations of a vacuum were added into the system to counterbalance the effect of the radiative damping—see, e.g., papers by Puthoff [13], Cole and Zou [14], as well as by Nieuwenhuizen [15], and references from these papers. However, first and foremost, the zero-point fluctuations are purely quantum effects. These kind of works are thus beyond the scope of the present paper devoted to purely classical description of microscopic systems. Second, the study by Nieuwenhuizen [15] (the latest out of the above three works) showed that this concept leads to the self-ionization of hydrogen atoms from states of relatively low (by absolute value) energy. Thus, this mixed quantum–classical concept actually does not explain the stability of all states of hydrogen atoms.

In the present paper, we apply the GHD to a charged Spherical Harmonic Oscillator (hereafter SHO). The SHO is important both fundamentally and practically. From the theoretical point of view, the SHO is one of the two fundamental microscopic systems (the other one being hydrogenic atoms), characterized by higher-than-geometrical symmetry (i.e., algebraic symmetry) and thus having conserved quantities beyond energy and the angular momentum. The algebraic symmetries of these two fundamental microscopic systems manifest classically by closed orbits and quantally by an "additional" degeneracy of their energy levels. From the practical point of view, the SHO is employed, e.g., in nuclear physics in nuclear shell models.

According to classical physics, the charged SHO is unstable with respect to radiating electromagnetic waves: it would lose its energy and end up in the state of zero energy. Of course, this is contrary to quantum mechanics, according to which the charged SHO has relatively stable stationary states. We show that, similarly to the Oks–Uzer results [8], the GHD—the purely classical formalism—allows the existence of, generally speaking, an infinite number of classical non-radiating states of the SHO.

2. Overview of the General Formalism and of Its Application to Hydrogenic Atoms

This overview is absolutely necessary for readers to facilitate their understanding of the new results (presented in Section 3) and of the conclusions (Section 4). The alternative would be to simply refer to our previous publications [8,10], but, in this case, readers would have to spend lots of time searching through our paper [8] and book [10]. Therefore, out of the respect to readers, we provide here excerpts from our two previous publications [8,10] as quotations (enclosed in the quotation marks).

From [8]:

"Dirac [1–3] considered a dynamical system of N degrees of freedom characterized by generalized coordinates q_n and velocities $v_n = dq_n/dt$, where $n = 1, 2, ..., N$. From the Lagrangian of the system

$$L = L(q_n, v_n) \tag{1}$$

momenta are defined as

$$p_n = \partial L/\partial v_n \tag{2}$$

From [10]:

"The quantities q_n, v_n, p_n can be varied by small amounts δq_n, δv_n, δp_n, respectively. The latter small quantities are of the order of ε and the variation should be worked to the accuracy of ε. As a result of the variation, the set of Equation (2) would not be satisfied any more. This is because their right side would differ from the corresponding left side by a quantity of the order of ε.

Further, Dirac made a distinction between two types of equations. One type is equations that do not hold after the variation, such as the set of Equation (2). Dirac called them "weak" equations. Below for weak equations, following Dirac, we use an equality sign $\cong$ different from the usual equality sign. Another type constitute equations that hold exactly even after the variation, such as Equation (1). Dirac called them "strong" equations. If quantities $\partial L/\partial v_n$ are not independent functions of velocities, it is possible to exclude velocities v_n from the set of Equation (2) and obtain one or several weak equations

$$\varphi(q, p) \cong 0 \tag{3}$$

containing only the sets of q and p (here and below we skip the suffix of quantities q and p)."

From [8]:

"In his formalism, Dirac [1–3] used the following complete system of independent Equations of the type (3):

$$\varphi_m(q, p) \cong 0, (m = 1, 2, ..., M) \quad (4)$$

Here the word "independent" means that neither of the φ's can be expressed as a linear combination of the other φ's with coefficients depending on q and p. The word "complete" means that any function of q and p, which would become zero with the allowance for Equation (4) and which would change by ε under the variation, should be a linear combination of the functions $\varphi_m(q, p)$ from Equation (4) with coefficients depending on q and p.

Finally, proceeding from the Lagrangian to a Hamiltonian, Dirac [1–3] obtained the following primary result:

$$H_g = H(q, p) + u_m \varphi_m(q, p) \quad (5)$$

(here and below, the summation over a twice repeated suffix is understood)."

From [10]:

"Equation (5) is a strong equation expressing a relation between the generalized Hamiltonian H_g and the conventional Hamiltonian H (q, p). Quantities u_m are coefficients to be determined."

From [8]:

"Generally, they are functions of q, v, and p; by using the set of Equation (2), they could be made functions of q and p. It should be emphasized that $H_g \cong H(q, p)$ would be only a weak equation - in distinction to Equation (5).

From Equation (5) it is seen that the Hamiltonian is not uniquely determined, because a linear combination of φ's may be added to it. Equation (4) are called constraints. The distinction between constraints (i.e., weak equations) and strong equations, described above, can be reformulated as follows.

Constraints must be employed in accordance to certain rules. Constraints can be added. Constraints can be multiplied by factors (depending on q and p), but only on the left side, so that these factors must not be used inside Poisson brackets.

If f is some function of q and p, then df/dt (i.e., a general equation of motion) in the Dirac's GHD is

$$df/dt \cong [f, H] + u_m[f, \varphi_m] \quad (6)$$

where [f, g] is the usual Poisson bracket. Substituting $\varphi_{m'}$ in Equation (6) instead of f and taking into account the set of Equation (4), one obtains:

$$[\varphi_{m'}, H] + u_m[\varphi_{m'}, \varphi_m] \cong 0. \quad (m' = 1, 2, ..., M) \quad (7)$$

these consistency conditions allow determining the coefficients u_m."

From [10]:

"It should be emphasized that the GHD was designed by Dirac specifically for applications to quantum field theory [3], that is, for the purpose totally different from the purpose of Oks-Uzer work [8]."

From [8]:

"The authors of paper [8] reformulated the GHD for atomic and molecular physics where many systems have a higher than geometrical symmetry and therefore possess additional integrals of the motion. Oks and Uzer [8] suggested using integrals of the motion as the constraints in the GHD.

In their general formalism, they considered a classical atomic or molecular system of N degrees of freedom, possessing M classical integrals of the motion A_m (q, p), m = 1, 2, ..., M. They wrote the generalized Hamiltonian in the form (see Equations (4) and (5)):

$$H_g = H(q, p) + u_m\{A_m(q, p) - A_{0m}\}, A_{0m} = \text{const.} \quad (8)$$

Here A_{0m} is the value of A_m (q, p) in a particular state of the motion, so that in this state

$$A_m(q, p) - A_{0m} \cong 0. \quad (9)$$

Since the quantities $A_m(q, p)$ are integrals of the motion, their Poisson bracket with H(q, p) vanishes and the consistency condition (7) reduces to the form

$$u_m[A_{m'}, A_m] \cong 0. \quad (m' = 1, 2, ..., M). \quad (10)$$

From [10]:

"The set of Equation (10) allows determining the coefficients u_m.

Specifically, for a hydrogenic atom of the nuclear charge Z, the integrals of the motion (other than the energy) are the angular momentum **L** = **r**□**p** and the Runge-Lenz vector (see, e.g., [16]) **A**(**r**,**p**) = {**r**p^2 − **p**(**r**·**p**)}/(μZe^2) − **r**/r, where μ is the reduced mass. Therefore, Oks and Uzer [8] presented the generalized Hamiltonian in the form:

$$H_g = p^2/(2\mu) - Ze^2/r + \mathbf{u}\times(\mathbf{r}\square\mathbf{p} - \mathbf{L}_0) + \mathbf{w}\times(\mathbf{A}(\mathbf{r},\mathbf{p}) - \mathbf{A}_0). \quad (11)$$

Here $\mathbf{L}_0$, $\mathbf{A}_0$, and the energy H_0 are connected by the well-known relation [16]:

$$L_0^2 = \mu Z^2e^4(A_0^2 - 1)/(2H_0). \quad (12)$$

From [8]:

"The consistency conditions [**r**□**p**, H_g] ≅ 0, [**A**(**r**,**p**), H_g] ≅ 0 resulted into the following equations for the unknown vector-coefficients **u** and **w**:

$$\mathbf{u}\square\mathbf{L}_0 + \mathbf{w}\square\mathbf{A}_0 \cong 0, \mathbf{u}\square\mathbf{A}_0 - 2\mathbf{w}\square\mathbf{A}_0H_0/(\mu Z^2e^4) \cong 0. \quad (13)$$

By using the consistency Equation (13), Oks and Uzer [8] reduced the number of yet unknown coefficients to just one, which they denoted as B. Of course, B was yet unknown function of energy H_0 in the particular state of the atom. Oks and Uzer [8] showed that in terms of B(H_0), the generalized Hamiltonian and the equations of the motion take the following form:

$$H_g = p^2/(2\mu) - Ze^2/r + 2B(H_0)H_0\{\mathbf{M}_0\cdot(\mathbf{r}\square\mathbf{p})/M_0^2 - (1 - \mathbf{A}_0\cdot\mathbf{A}(\mathbf{r},\mathbf{p}))/(1 - A_0^2)\} \quad (14)$$

$$d\mathbf{r}/dt = \{1 + B(H_0)\}\mathbf{p}/\mu \quad (15)$$

$$d\mathbf{p}/dt = -\{1 + B(H_0)\}Ze^2\mathbf{r}/r^3$$

From [10]:

"The Equation of the motion (15) differ from their conventional form only by the factor {1 + B(H_0, A_0)}. Therefore, the transformation of the time

$$t' = \{1 + B(H_0)\}t \quad (16)$$

the equations of the motion with respect to the new time t′ would be formally brought back to their conventional form.

Thus the authors of paper [8] came to the following central point. In the above generalized formalism, the trajectory of the atomic electron remains the same as in the conventional formalism. However, the generalized period T_g and the generalized frequency ω_g differ from their conventional values T_0 and ω_0 as follows:

$$T_g - T_0/|1 + B(H_0)| \quad (17)$$

$$\omega_g = \omega_0|1 + B(H_0)| = |1 + B(H_0)||2H_0|^{3/2}/D^{1/2}, D \cong \mu Z^2e^4 \quad (18)$$

(in Equation (18), the explicit expression for the Kepler frequency ω_0 has been used)."

From [8]:

"Equation (18) clearly demonstrates that the generalized formalism allows the existence of such state (or states) of the motion, where $\omega_g = 0$ despite $H_0 \neq 0$ (the conventional formalism allows to be $\omega_0 = |2H_0|^{3/2}/D^{1/2} = 0$ only for $H_0 = 0$). This is a state (or states) where $B(H_0) = -1$. Therefore, such state or states would not emit the electromagnetic radiation, would not lose energy for the radiation, and would thus constitute stable states of the classical atom."

The authors of paper [8] showed that there is infinite number of the energies of the classical stationary states—just as in the corresponding quantum solution.

From [10]:

"Oks and Uzer [8] pointed out that in a classical non-radiating stable state, one has $d\mathbf{r}/dt = d\mathbf{p}/dt = 0$, so that $\mathbf{r}(t) = \mathbf{r}_0$ and $\mathbf{p}(t) = \mathbf{p}_0$, where $\mathbf{r}_0$ and $\mathbf{p}_0$ are some vector constants. Thus, the electron is motionless, but its momentum differs from zero. This is not surprising: the momentum $\mathbf{p}$ is a more complex physical quantity than the velocity $\mathbf{v} \equiv d\mathbf{r}/dt$. For example, for a charge in an electromagnetic field characterized by a vector-potential $\mathbf{A}$, it is also possible to have $\mathbf{v} = [\mathbf{p} - e\mathbf{A}/(mc)]/m = 0$ while $\mathbf{p} = e\mathbf{A}/(mc) \neq 0$ [17].

It is also very important to emphasize that the physics behind such classical non-radiating states is a new kind of time-dilation expressed by Equation (16): a non-Einsteinian time-dilation, as pointed out in book [10]. The closer the energy of the system to the energy of the classical non-radiating state, the more dilates the time. At the classical non-radiating state, the time gets dilated infinitely, so that the frequency ω_g in Equation (18) vanishes and so does the radiation."

3. New Results

We consider a charged Spherical Harmonic Oscillator (SHO). The "conventional" conserved quantities are the energy E and the angular momentum vector **M**, the conservation of the latter following from the geometrical (spherical) symmetry of this system. It is well-known that the SHO also possesses another set of conserved quantities, whose conservation is the consequence of the higher-than-geometric (algebraic) symmetry:

$$I_{mn} = p_m p_n/\mu + k x_m x_n,\ m = 1, 2, 3,\ n = 1, 2, 3 \tag{19}$$

Here, p_m and x_m are the Cartesian components of the momentum $\mathbf{p}$ and of the radius-vector $\mathbf{r}$, respectively; μ is the mass of the SHO. Obviously, $I_{nm} = I_{mn}$, so that there are generally only six independent conserved quantities I_{mn}. The unperturbed Hamiltonian H can be actually expressed via some of the conserved quantities from Equation (19) as follows:

$$H = (I_{11} + I_{22} + I_{33})/2 \tag{20}$$

It is well known that the motion is limited to a plane. We choose the x_3-axis (the z-axis) perpendicular to the orbital plane. Then, the dynamical variables are x_1, p_1, x_2, p_2.

In order to study whether classical non-radiative states of the SHO are possible, it should be sufficient to consider the generalized Hamiltonian H_g, which differs from H only by the addition of the constraints corresponding to the conserved quantities responsible for the algebraic symmetry, i.e., the conserved quantities from Equation (19), but only those of them that are relevant to the motion in the orbital plane:

$$H_g = (I_{11} + I_{22})/2 + B_{11}(E)\,(I_{11} - I_{11,0}) + B_{22}(E)\,(I_{22} - I_{22,0}) + B_{12}(E)\,(I_{12} - I_{12,0}) \tag{21}$$

where $I_{mn,0}$ are the values of these conserved quantities in the particular state of the system; E is the energy of the system in a particular state of the motion.

The conserved quantities I_{mn} "commute" with each other: the Poisson bracket of any two of them vanishes. Therefore, the consistency conditions from Equation (10) in this case reduce to equating to zero the Poisson brackets of the components of the angular momentum **M** with the second term in the right side of Equation (21):

$$[M_i, a_{mn} I_{mn}] = [e_{ijq}x_j p_q, B_{mn}(p_m p_n/\mu + k x_m x_n)] = 0 \tag{22}$$

where e_{ijq} is the Levi-Civita symbol.

The calculations of the Poisson brackets from Equation (22), with the subsequent substitution of I_{mn} by $I_{mn,0}$ (as required by the GHD), lead to the following equations:

$$B_{22}I_{12,0} = B_{12}I_{22,0} \tag{23}$$

$$B_{12}I_{11,0} = B_{11}I_{12,0} \tag{24}$$

From Equation (19), it is obvious that the quantities I_{11} and I_{22} are non-negatively defined.

For definiteness, we assume that $I_{11,0}$ differs from zero, i.e., $I_{11,0} > 0$. Then, from Equations (23) and (24), it is easy to obtain

$$B_{12} = B_{11}I_{12,0}/I_{11,0}, \quad B_{22} = B_{11}I_{22,0}/I_{11,0} \tag{25}$$

Thus, the consistency conditions help reduce the unknown coefficients in the generalized Hamiltonian H_g from three to one, so that H_g can be represented in the form:

$$H_g = (I_{11} + I_{22})/2 + B_{11}(E)\,\{(I_{11} - I_{11,0}) + I_{22,0}/I_{11,0}\,(I_{22} - I_{22,0}) + I_{12,0}/I_{11,0}\,(I_{12} - I_{12,0})\} \tag{26}$$

Based on the Hamiltonian H_g from Equation (26) and using $dx_i/dt = \partial H_g/\partial p_i$, $dp_i/dt = -\partial H_g/\partial x_i$, we find the following equations of motion:

$$dx_1/dt = \{(1+2B_{11})p_1 + (B_{11}I_{12,0}/I_{11,0})p_2,\ dx_2/dt = (1+2\,B_{11}I_{22,0}/I_{11,0})p_2 + (B_{11}I_{12,0}/I_{11,0})p_1\}/\mu \tag{27}$$

$$dp_1/dt = -k\{(1+2B_{11})x_1 + (B_{11}I_{12,0}/I_{11,0})x_2,\ dp_2/dt = (1+2\,B_{11}I_{22,0}/I_{11,0})p_2 + (B_{11}I_{12,0}/I_{11,0})x_1\} \tag{28}$$

By differentiation of Equation (27) with respect to time and substituting Equation (28) into the outcome, we obtain the following system of equations:

$$d^2x_1/dt^2 = -\omega_0^2\{[(1+2B_{11})^2 + (B_{11}I_{12,0}/I_{11,0})^2]x_1 + 2(B_{11}I_{12,0}/I_{11,0})[1 + B_{11}(1+ I_{22,0}/I_{11,0})]x_2\}, \tag{29}$$

$$d^2x_2/dt^2 = -\omega_0^2\{[(1+2B_{11}I_{22,0}/I_{11,0})^2 + (B_{11}I_{12,0}/I_{11,0})^2]x_2 + 2(B_{11}I_{12,0}/I_{11,0})[1 + B_{11}(1+I_{22,0}/I_{11,0})]x_2\}$$

where

$$\omega_0 = (k/\mu)^{1/2} \tag{30}$$

is the "unperturbed" frequency of the oscillator.

We seek a solution of system (29) in the form:

$$x_1 = \exp(i\omega_g t),\ x_2 = \alpha \exp(i\omega_g t),\ \alpha = \text{const} \tag{31}$$

here, ω_g is the (yet unknown) generalized frequency of the oscillator.

Substituting x_1 and x_2 from Equation (31) into the first equation in Formula (29), we obtain:

$$\omega^2/\omega_0^2 = (1+2B_{11})^2 + (B_{11}I_{12,0}/I_{11,0})^2 + 2\alpha(B_{11}I_{12,0}/I_{11,0})[1 + 2B_{11}(1 + I_{22,0}/I_{11,0})] \tag{32}$$

Substituting x_1 and x_2 from Equation (31) into the second equation in Formula (29), we obtain:

$$\omega_g^2/\omega_0^2 = (1+2B_{11}I_{22,0}/I_{11,0})^2 + (B_{11}I_{12,0}/I_{11,0})^2 + (2/\alpha)(B_{11}I_{12,0}/I_{11,0})[1 + 2B_{11}(1+ I_{22,0}/I_{11,0})] \quad (33)$$

For Equations (32) and (33), which have the same left sides, to be compatible with each other, their right sides should also be equal to each other. By equating the right sides of Equations (32) and (34), after some simplifications, we find that the parameter α must satisfy the following quadratic equation:

$$\alpha^2 - 2\gamma\alpha - 1 = 0, \gamma = (I_{22,0} - I_{11,0})/I_{12,0} \quad (34)$$

The two solutions of Equation (34) are

$$\alpha_\pm = \gamma \pm (\gamma^2 + 1)^{1/2} \quad (35)$$

Obviously, $\alpha_+ > 0$ while $\alpha_- < 0$. Physically, these two solutions correspond to the two opposite directions of the revolution along the orbit (see Equation (31)).

Equation (33) can be represented in a more explicit form:

$$\omega_g^2/\omega_0^2 = (4 + 2\alpha_\pm\varepsilon\delta + \delta^2)B_{11}^2 + 2(2 + \alpha_\pm\delta)B_{11} + 1 \quad (36)$$

where we temporarily introduce the following notation:

$$\varepsilon = (1 + I_{22,0}/I_{11,0}), \quad \delta = I_{12,0}/I_{11,0} \quad (37)$$

Using Equation (35), it is easy to find out that

$$4 + 2\alpha_\pm\varepsilon\delta + \delta^2 = (2 + \alpha_\pm\delta)^2 \quad (38)$$

so that Equation (35) simplifies to

$$\omega_g^2/\omega_0^2 = [(2 + \alpha_\pm\delta)B_{11} + 1]^2 \quad (39)$$

which is equivalent to the following:

$$\omega_g/\omega_0 = |(2 + \alpha_\pm\delta)B_{11} + 1| \quad (40)$$

Coming back to the original notations, we rewrite Equation (40) in the form

$$\omega_g/\omega_0 = |\{1 + I_{22,0}/I_{11,0} \pm [(I_{22,0}/I_{11,0} - 1)^2 + I_{12,0}^2/I_{11,0}^2]^{1/2}\}B_{11}(E) + 1| \quad (41)$$

where we have restored the argument E of the coefficient $B_{11}(E)$. It is seen that, for each direction of the revolution of the charged particle in the orbital plane, there is a value of $B_{11}(E)$, for which the generalized frequency is ω_g vanishes and so is the radiation. These non-radiating (stationary) states correspond explicitly to the following values, $B_{11+}(E_{st})$ and $B_{11-}(E_{st})$ of $B_{11}(E)$, where the subscript "st" stands for "stationary":

$$B_{11+}(E_{st}) = -1/\{1 + I_{22,0}/I_{11,0} + [(I_{22,0}/I_{11,0} - 1)^2 + I_{12,0}^2/I_{11,0}^2]^{1/2}\} \quad (42)$$

for $\alpha = \alpha_+$ and

$$B_{11-}(E_{st}) = -1/\{1 + I_{22,0}/I_{11,0} - [(I_{22,0}/I_{11,0} - 1)^2 + I_{12,0}^2/I_{11,0}^2]^{1/2}\} \quad (43)$$

for $\alpha = \alpha_-$. Remember that $\alpha_\pm(\gamma)$ is given by Equation (35), where $\gamma = (I_{22,0} - I_{11,0})/I_{12,0}$.

Figure 1 shows a three-dimensional plot of B_{11+} (denoted in the plot for brevity as B_+) versus $I_{22,0}/I_{11,0}$ (denoted in the plot as C) and $I_{12,0}/I_{11,0}$ (denoted in the plot as D).

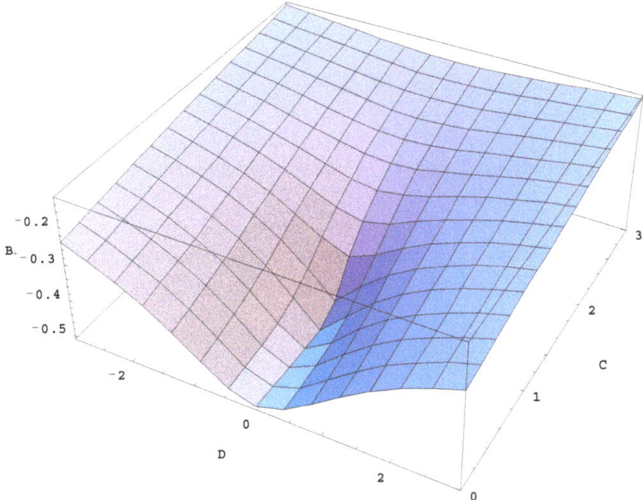

Figure 1. Three-dimensional plot of B_{11+} (denoted in the plot for brevity as B_+) from Equation (42) versus $I_{22,0}/I_{11,0}$ (denoted in the plot as C) and $I_{12,0}/I_{11,0}$ (denoted in the plot as D).

Figure 2 shows a three-dimensional plot of B_{11-} (denoted in the plot for brevity as B_-) versus $I_{22,0}/I_{11,0}$ (denoted in the plot as C) and $I_{12,0}/I_{11,0}$ (denoted in the plot as D).

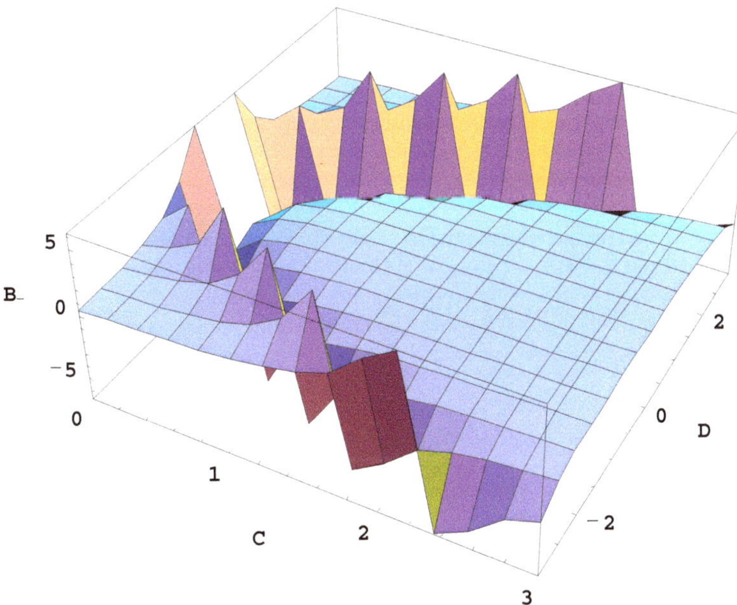

Figure 2. Three-dimensional plot of B_{11-} (denoted in the plot for brevity as B_-) from Equation (43) versus $I_{22,0}/I_{11,0}$ (denoted in the plot as C) and $I_{12,0}/I_{11,0}$ (denoted in the plot as D).

Here is an intermediate summary of the above results. By employing the GHD, we have proven the existence of the classical non-radiating states of the charged spherical harmonic oscillator—similarly to the corresponding results of paper [8] for hydrogenic atoms.

Physically, this is the manifestation of a non-Einsteinian time-dilation. Time dilates more and more as the energy of the system becomes closer and closer to the energy of the classical non-radiating state:

$$t' = |\{1 + I_{22,0}/I_{11,0} \pm [(I_{22,0}/I_{11,0} - 1)^2 + I_{12,0}{}^2/I_{11,0}{}^2]^{1/2}\}B_{11}(E) + 1|t \quad (44)$$

At the classical non-radiating state, the time gets dilated infinitely. As a result, the frequency of the revolution along an elliptical orbit ω_g in Equation (41) vanishes—consequently, the radiation also vanishes.

In the important particular case of $I_{22,0} = I_{11,0}$, corresponding to the circular orbits, the above formulas can be simplified as follows. In this case, from Equation (34) follows $\gamma = 0$, so that from Equation (35) we get $\alpha_+ = 1$ and $\alpha_- = -1$, as it should be for the circular orbits (see Equation (31)). Then Equation (36) simplifies to

$$\omega_g/\omega_0 = |(2 \pm |I_{12,0}|/I_{11,0})B_{11}(E) + 1| \quad (45)$$

where the plus sign corresponds to $\alpha = 1$ and the minus sign corresponds to $\alpha = -1$.

It is seen that, in this particular case, the generalized frequency ω_g vanishes (and so does the radiation) at the following values of $B_{11}(E)$:

$$B_{11+}(E_{st}) = -1/(2 + |I_{12,0}|/I_{11,0}) \text{ for } \alpha = 1 \quad (46)$$

and

$$B_{11-}(E_{st}) = -1/(2 - |I_{12,0}|/I_{11,0}) \text{ for } \alpha = -1 \quad (47)$$

Obviously, Equation (47) is valid, except if $|I_{12,0}|/I_{11,0} = 2$. In the exceptional case, Equation (45) yields a trivial result: $\omega_g = \omega_0$.

The primary result for the circular orbits is non-trivial. Namely, there are classical non-radiating states, corresponding to $B_{11}(E) = B_{11+}(E_{st})$ for $\alpha = 1$ or $B_{11}(E) = B_{11-}(E_{st})$ for $\alpha = -1$.

Thus, for each direction of the revolution of the charged particle in the orbital plane, there is one value of $B_{11}(E)$—given by Equations (42) and (43) in the general case of the elliptical orbits or by Equations (46) and (47) for the particular case of the circular orbits—for which the radiation vanishes. The fact that, for each direction of the revolution, there is only one value of $B_{11}(E)$, does not mean that there is only one classical stationary state. Indeed, if the dependence of B_{11} on the energy E is oscillatory (with the amplitude greater than or equal to the absolute value of the right side of Equation (42) for $\alpha = \alpha_+$, or with the amplitude greater than or equal to the absolute value of the right side of Equation (43) for $\alpha = \alpha_-$), then there would be an infinite number of the energies of the classical stationary states E_{st}—just as in the corresponding quantum solution.

Here is an example, illustrating the statement from the previous sentence for the case of circular orbits—for the subcase of $\alpha = 1$ chosen for definiteness. Let us consider the following dependence of B_{11+} on the energy E:

$$B_{11+}(E) = -|\cos[\pi(E - C)/(E_{st,0} - C)]|/(2 + |I_{12,0}|/I_{11,0}) \quad (48)$$

where $E_{st,0}$ is the energy of the lowest non-radiating state (the ground state) and both E and $E_{st,0}$ are measured in units of $\hbar\omega_0$. In Equation (48), C is a constant, which is an analog of the Maslov index [18], which, for spherically symmetric potentials, is equal to 1/2 (see, e.g., the textbook [19]). With C = 1/2, Equation (48) takes the form

$$B_{11+}(E) = -|\cos[\pi(E - 1/2)/(E_{st,0} - 1/2)]|/(2 + |I_{12,0}|/I_{11,0}) \quad (49)$$

From Equation (49), it is easy to find out that $E = E_{st,n}$, where

$$E_{st,n} - 1/2 = (n + 1)(E_{st,0} - 1/2), \quad n = 0, 1, 2, \ldots, \tag{50}$$

where the quantity B_{11+} satisfies Equation (46), so that the sequence of values $E_{st,n}$ from Equation (50) is the sequence of the energies of the classical non-radiating stationary states. More explicitly,

$$E_{st,n} = (n + 1)E_{st,0} - n/2 \tag{51}$$

If $E_{st,0} = 3/2$, then the sequence of values $E_{st,n}$ from Equation (51) would coincide with the corresponding quantum results.

4. Conclusions

We extended the applications of the GHD to a charged Spherical Harmonic Oscillator (SHO). We demonstrated that, by using the higher-than-geometrical symmetry (i.e., the algebraic symmetry) of the SHO and the corresponding additional conserved quantities, it is possible to obtain classical non-radiating (stationary) states of the SHO. Generally, there is an infinite number of such states of the SHO—just as was the case for hydrogenic atoms, as was shown in paper [8]. Both the existence of the classical stationary states of the SHO and the infinite number of such states are consistent with the corresponding quantum results.

Physically, the existence of the classical stationary states is the manifestation of a non-Einsteinian time-dilation. Time dilates more and more as the energy of the system becomes closer and closer to the energy of the classical non-radiating state.

It should be emphasized that we obtained the above new results from first principles. We did not use any quantization postulates or any input from experiments.

It is worth mentioning that the SHO and hydrogenic atoms are not the only microscopic systems that can be successfully treated by the GHD. Indeed, all classical systems of N degrees of freedom have the algebraic symmetries O_{N+1} and SU_N, and this does not depend on the functional form of the Hamiltonian. In particular, all classical spherically symmetric potentials have algebraic symmetries, namely tO_4 and SU_3; they possess an additional vector integral of the motion, while the quantal counterpart-operator does not exist [20–22]. (This fact was employed in paper [9], where the authors successfully applied the GHD to a modified Coulomb potential.) This offers possibilities that are absent in quantum mechanics, as noted in paper [8].

Since there are lots of classical systems possessing an algebraic symmetry and, therefore, having additional integrals of the motion, as mentioned in the previous paragraph, it should be obvious that the classical systems studied in papers [8,9] and in the present paper do not constitute a comprehensive list. For example, another fundamental physical system—an electron in the field of two stationary nuclei—is a good candidate to be treated by GHD. Indeed, this system has an additional integral of the motion—the projection of the super-generalized Runge–Lenz vector on the internuclear axis, the latter vector being derived in paper [23].

Funding: This research received no external funding.

Conflicts of Interest: The author declares no conflict of interest.

References

1. Dirac, P.A.M. Generalized Hamiltonian dynamics. *Canad. J. Math.* **1950**, *2*, 129–148. [CrossRef]
2. Dirac, P.A.M. Generalized Hamiltonian dynamics. *Proc. R. Soc. A* **1958**, *246*, 326–332. [CrossRef]
3. Dirac, P.A.M. *Lectures on Quantum Mechanics*; Academic: New York, NY, USA, 1964; Reprinted by Dover Publications, 2001.
4. Sergi, A. Generalized bracket formulation of constrained dynamics in phase space. *Phys. Rev. E* **2004**, *69*, 21109. [CrossRef] [PubMed]

5. Sergi, A. Phase space flows for non-Hamiltonian systems with constraints. *Phys. Rev. E* **2005**, *72*, 31104. [CrossRef] [PubMed]
6. Sergi, A. Non-Hamiltonian commutators in quantum mechanics. *Phys. Rev. E* **2005**, *72*, 66125. [CrossRef] [PubMed]
7. Sergi, A. Statistical mechanics of quantum-classical systems with holonomic constraints. *J. Chem. Phys.* **2006**, *124*, 24110. [CrossRef] [PubMed]
8. Oks, E.; Uzer, T. Application of Dirac's generalized Hamiltonian dynamics to atomic and molecular systems. *J. Phys. B At. Mol. Opt. Phys* **2002**, *35*, 165–173. [CrossRef]
9. Camarena, A.; Oks, E. Application of the generalized Hamiltonian dynamics to a modified coulomb potential. *Int. Rev. At. Mol. Phys.* **2010**, *1*, 143–160.
10. Oks, E. *Breaking Paradigms in Atomic and Molecular Physics*; World Scientific: Singapore, 2015.
11. Goedecke, G.H. Classically radiationless motions and possible implications for quantum theory. *Phys. Rev.* **1964**, *135*, B281–B288. [CrossRef]
12. Raju, C.K. The Electrodynamic 2-body problem and the origin of quantum mechanics. *Found. Phys.* **2004**, *34*, 937–962. [CrossRef]
13. Puthoff, H.E. Ground state of hydrogen as a zero-point-fluctuation-determined state. *Phys. Rev. D* **1987**, *35*, 3266–3269. [CrossRef] [PubMed]
14. Cole, C.D.; Zou, Y. Quantum mechanical ground state of hydrogen obtained from classical electrodynamics. *Phys. Lett. A* **2003**, *317*, 14–20. [CrossRef]
15. Nieuwenhuizen, T.M. On the stability of classical orbits of the hydrogen ground state in stochastic electrodynamics. *Entropy* **2006**, *18*, 135. [CrossRef]
16. Landau, L.D.; Lifshitz, E.M. *Mechanics*; Pergamon Press: Oxford, UK, 1960.
17. Landau, L.D.; Lifshitz, E.M. *Classical Theory of Fields*; Pergamon Press: Oxford, UK, 1960.
18. Mukunda, N. Dynamical symmetries and classical mechanics. *Phys. Rev.* **1967**, *155*, 1383–1386. [CrossRef]
19. Bacry, H.; Ruegg, H.; Souriau, J.M. Dynamical groups and spherical potentials in classical mechanics. *Commun. Math. Phys.* **1966**, *3*, 323–333. [CrossRef]
20. Fradkin, D.M. Existence of the dynamic symmetries O4 and SU3 for all classical central potential problems. *Progr. Theor. Phys.* **1967**, *37*, 798–812. [CrossRef]
21. Maslov, V.P.; Fedoriuk, M.V. *Semi-Classical Approximation in Quantum Mechanics*; Springer: Berlin, Germany, 1981.
22. Landau, L.D.; Lifshitz, E.M. *Quantum Mechanics*; Pergamon Press: Oxford, UK, 1965.
23. Kryukov, N.; Oks, E. Super-generalized Runge-Lenz vector in the problem of two coulomb or newton centers. *Phys. Rev. A* **2012**, *85*, 54503. [CrossRef]

MDPI
St. Alban-Anlage 66
4052 Basel
Switzerland
Tel. +41 61 683 77 34
Fax +41 61 302 89 18
www.mdpi.com

Symmetry Editorial Office
E-mail: symmetry@mdpi.com
www.mdpi.com/journal/symmetry

www.ingramcontent.com/pod-product-compliance
Lightning Source LLC
LaVergne TN
LVHW070616170726
843515LV00004B/94

* 9 7 8 3 0 3 6 5 5 4 4 3 3 *